Volume I

Molecular Electronics and Molecular Electronic Devices

VOLUME I

MOLECULAR ELECTRONICS AND MOLECULAR ELECTRONIC DEVICES

EDITED BY

KRISTOF SIENICKI

Faculty of Applied Physics and Mathematics
Technical University of Gdańsk
Gdańsk, Poland

CRC Press
Boca Raton   Ann Arbor   London   Tokyo

Library of Congress Cataloging-in-Publication Data

Molecular electronics and molecular electronic devices / editor,
Kristof Sienicki.
p. cm.
Includes bibliographical references and index.
ISBN 0-8493-8061-8 (v. 1)
1. Molecular electronics. I. Sienicki, Kristof.
TK7874.8.M654 1992
621.381--dc20 92-25628
CIP

This book represents information obtained from authentic and highly regarded sources. Reprinted material is quoted with permission, and sources are indicated. A wide variety of references are listed. Every reasonable effort has been made to give reliable data and information, but the author and the publisher cannot assume responsibility for the validity of all materials or for the consequences of their use.

Direct all inquiries to CRC Press, Inc., 2000 Corporate Blvd., N.W., Boca Raton, Florida, 33431.

International Standard Book Number 0-8493-8061-8

Library of Congress Card Number 92-25628

Printed in the United States of America 1 2 3 4 5 6 7 8 9 0

Printed on acid-free paper

PREFACE

The emerging field of molecular electronics and molecular electronic devices presents new challenges and opportunities for scientists and researchers working in physics, chemistry, biology, and computer science. The synthesis of these subjects and the hope of developing molecular structures capable of performing computations on a molecular level are especially captivating. This book series entitled *Molecular Electronics and Molecular Electronic Devices* addresses these two concerns and reflects the state of the art in this field. Experimental and theoretical aspects of the field are reviewed by researchers actively working in physics, chemistry, computer science, and various biological subjects. This series will make a valuable contribution by helping to unify scientists concerned with molecular electronics and molecular electronic devices.

Although the manuscripts published are specially commissioned, researchers are welcome to make suggestions to the Editor.

Kristof Sienicki
October, 1992

The Editor

Kristof Sienicki, Ph.D., is at the Institute of Technical Physics and Applied Mathematics, Technical University of Gdańsk (Poland), where he has been working in molecular physics and molecular electronics since 1980. His past and present research interests lie in the theoretical, experimental, and numerical studies of the photophysics of aromatic polymers and molecules, energy and electron transport, Langmuir-Blodgett layers, and neural networks.

Dr. Sienicki obtained his Ph.D. in physics from the Institute of Physics (University of Gdańsk) in 1985. As a postdoctoral fellow, researcher, and visiting professor, he spent 7 years at various universities in the United States and Canada. He has published more than 50 research papers, and he has presented a number of lectures at international meetings as well as guest lectures at universities. He is currently teaching a course in molecular electronics.

Contributors

D. S. Chernavskii
P.N. Lebedev Physical Institute
Russian Academy of Sciences

Héctor Cruz
Departmento de Fisica Fundamental y Experimental
Universidad de La Laguna
La Laguna, Tenerife, Spain

Aleksander Herman
Department of Chemistry
Technical University of Gdańsk
Gdańsk, Poland

Masao Kaneko
The Institute of Physical and Chemical Research
Wako, Saitama, Japan

Robert W. Keyes
IBM Research Division
Yorktown Heights, New York

V. I. Krinsky
P.N. Lebedev Physical Institute
Russian Academy of Sciences
Moscow, Russia

Ru-Juang Lin
The Institute of Physical and Chemical Research
Wako, Saitama, Japan

J. Gonzalo Muga
Departamento de Fisica Fundamental y Experimental
Universidad de La Laguna
La Laguna, Tenerife, Spain

N. G. Rambidi
International Research Institute for Management Sciences
Moscow, Russia

Rafael Sala
Departamento de Fisica Fundamental y Experimental
Universidad dc La Laguna
La Laguna, Tenerife, Spain

Shammai Speiser
Department of Chemistry
Technion-Israel Institute of Technology
Haifa, Israel

ADVISORY BOARD

CONTENTS

"No progress is possible without the endeavours of people, and one cannot stop people from endeavours."

(Source unknown)

Chapter 1

ELECTRONICS IN LARGE SYSTEMS

Robert W. Keyes

TABLE OF CONTENTS

0-8493-8061-8/93/$0.00 + $.50

I. INTRODUCTION

Artificial information-processing systems handle information in the form of the value of some physical variable. Electrical quantities have long been used to represent information because of the ease with which they can be transmitted through wires to other places where the information may be needed. However, other physical quantities have found use in technology. Much use was once made in machinery of cams and levers, for which position can be transmitted through short distances by shafts or rods. These ingredients were prominent in the early machines of Leibnitz and Babbage, for example.[1] Recent developments have demonstrated that light waves traveling through optical fibers offer advantages in the transmission of information over long distances. The pressure in a fluid is useful because it is easily transmitted a substantial distance through tubing.

Another advantage of electrical signals is the ease with which conversion between electrical and other representations of information can be achieved. Photodiodes and light-emitting diodes readily convert information between electrical and optical forms. The position of switches is especially suitable when a mechanical interface with humans is wanted. Information in the form of holes in cards or tapes was once widely used; electrical contact could be established through the holes to convert the information into electrical signals. The direction of magnetization of a body of ferromagnetic material can be controlled and sensed by electromagnetic induction.

Physical variables that represent information act on devices to produce some response, often an output variable of similar type that can be transmitted, in turn, to another device. While systems that perform such functions have existed for many decades, for example, mechanical adding and tabulating machines, they did not develop into a major industry until the advent of solid-state electronics. The later event was triggered by the invention of a solid-state device, the transistor, around the middle of the present century. The potential of solid-state devices was quickly recognized, large investments were attracted, and many inventions followed.

The development of electronics since the invention of the transistor has been marked by miniaturization. The functions that were once performed by vacuum tubes are now reproduced in devices that have volumes many orders of magnitude smaller and that require the finest UV optics or electron beams for their fabrication. The forces driving this tremendous miniaturization are cost, reliability, and power. Speed of operation has been a welcome byproduct of miniaturization. Huge decreases in cost and in power dissipation and large increases in reliability have made it possible to construct systems with enormous numbers of electronic components. To illustrate miniaturization, the area required to represent 1 bit of information in various memory and storage technologies is shown in Figure 1. The area per bit has been decreased by a factor of about 10^5 in the period covered. This large factor has been reflected in an enormous decrease in the cost per bit of information storage systems.

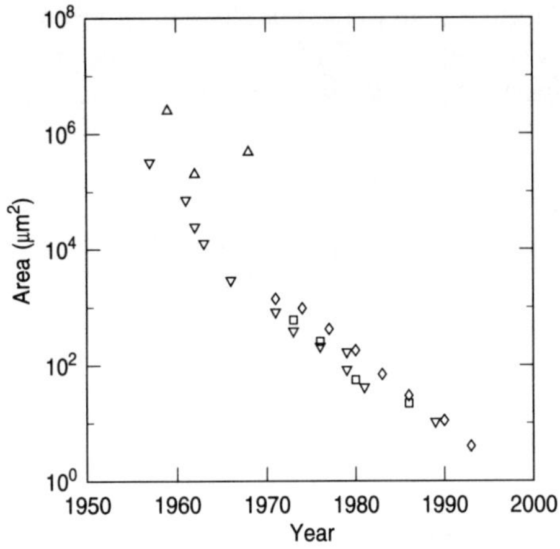

FIGURE 1. The progress of miniaturization: the area used to store a bit in various technologies as a function of time. Disk files: ▽; magnetic bubbles: □; thin-film memory: △; silicon DRAM: ◇.

The low cost accruing to miniaturization depends on the ability of fabrication tools to handle and process a great many devices simultaneously. The smaller the device and the smaller the distance between them, the larger the number that can be squeezed onto a semiconductor substrate, a wafer. Therefore, devices are packed as closely together as possible in these large systems. The silicon wafers that are handled in the fabrication of 4-Mb semiconductor memory chips may contain 200 chips, a total of about 10^9 bit of information on one wafer. The cost of producing the wafer is divided among all of these bits.

II. DEVICES IN SYSTEMS

A. SYSTEM REQUIREMENTS

The great power of electronic information handling derives from the fact that computational systems can bring a great many components together to make the performance of very complex functions possible. The large electronic systems that are constructed from many devices are computers or special purpose modifications of them. The devices in computers are organized into logic circuits made of a few transistors that can perform an elementary logic function such as an AND or a NOR.

Causing a large number of electronic components to work together to achieve some desired ends requires a large amount of communication between the components. Because the basic component — the transistor — is an electrical device, this communication is carried out by a dense network of electrical connections — wires.

Providing a large amount of communication between devices implies that the logic circuits must have means of sending signals on the network of wires. The wires on chips are short and can be regarded as an additional capacitance to be charged by the current that is controlled by a transistor. Their length and capacitance have decreased with the miniaturization of features on chips. Longer wires connect chips to one another and must be designed to avoid reflections that can interfere with signals. The current used to charge the wires has remained high in high-performance systems, the lowered capacitance enabling faster operation at the expense of increasing density of dissipation of electrical power to heat.

In large systems the cost factors that drive miniaturization, integration, and high density of components on chips apply to the substrates on which the chips are mounted, called modules, boards, or cards. The colloquial reference to area on chips and on chip substrates as "real estate" is essentially a sign of the high esteem in which surface area is held. In addition, placing the chips close together reduces the time for signals to travel through a system and assists system speed. Further, space in buildings is valuable in itself, and the cost of the volume occupied by electronic equipment is an incentive for the pursuit of miniaturization and dense packing. Additional motivation toward high density is provided by applications for mobile and portable systems.

B. SYSTEM ENVIRONMENT

The above factors create an unusual environment in large systems, in which logic circuits must operate, transmit, and receive signals. The close packing of devices, wires, and chips allows elements to influence their neighbors. Electromagnetic signals on the wires can induce voltages and currents in neighboring wires. The heat created in one place affects the temperature in nearby places. Device characteristics depend on temperature, and the temperature at which a device operates is not precisely known and may vary with time. The system must function correctly in spite of these hostile aspects of the environment.

The unavoidable imperfections of manufactured devices form another obstacle that must be overcome to achieve correct functioning of a system. The assembly of very large numbers of devices into a system is only possible because the cost of a device is very small. The low cost is achieved by mass fabrication, and precludes devoting attention and effort to individual devices. Also, there are factors that are not under the control of process technologies, such as the precise location of atoms of dopants, impurities, and crystalline defects. The fluxes of electrical and thermal currents to which devices are exposed can alter them and are another source of variability. In short, circuit and device designers must plan for a range of device characteristics.

Of course, such uncertainties in signals and in the entities that must receive and interpret them are not unique to computing systems. Electrical noise has been a nuisance since the invention of radio. All of us have suffered through poor telephone connections. Reflections from passing aircraft distort signals received by television antennas. The most satisfactory approach to reliable communication in a noisy environment is the representation of information in digital form. Digital means that there are just a small number of possible symbols; the recipient only has to determine which was transmitted. An early example, devised long ago for telegraphy, is the Morse code which was also adopted by and still is sometimes used in radio communication. Digital information was not invented for electrical technology: alphabets and representations of the integers are digital. The abacus exemplifies principles that underly modern electronic computation. Each bead of the abacus has two meaningful positions, corresponding to the binary system of numbers. Binary numbers can be put into one-to-one correspondence with the more familiar alphabetic characters and decimal digits to interface with people. Binary representation is used in all electronic computation and in information storage in computers. It is also widely used in communication and in recording for other purposes. An immunity to much of the effect of noise is obtained through the need to only decide between two alternatives. In terms of the abacus the establishment of two different fixed positions eliminates the uncertainty, the noise, that would be involved in trying to make a quantitative estimate of the position of a bead. Which way has the bead been pushed?

The information in Morse code is carried by the duration of a pulse. The presence or absence of a signal, a pulse of current or voltage or the intensity of a beam, is used in computers. The amount of information in one pulse is called a bit. Another advantage of digital representation for purposes of computation in that arbitrary accuracy is possible by using enough bits. Systems often offer a user the option of using "double precision", representation of a number by twice as many bits, if the default is insufficient. Integers of many thousands of bits are handled in the search for large primes.

The recipient of a binary digit must have some knowledge of what to expect, some standard with which a received pulse can be compared. Optimally, the values of digits are derived from a standard known to the receiver. A manufactured system must be designed with some knowledge of what the bits will be like. The detector at the end of an optical fiber in an amplitude modulated system interprets bits by comparison with a threshold determined by the number of photons that are expected to represent a pulse. The electrical pulses in a computer are derived by connection of an output to a power supply that is distributed throughout the system. The power supply is available at receiving devices for comparison with incoming signals.

Therefore, a receiver must distinguish between two possible values of a binary digit in the systems considered. A logic circuit determines an output bit from its received bits and produces a standardized signal to transmit. The idealized digital response is shown in Figure 2. This would apply to the abacus as well as to modern electronic signals. In the case of the abacus the horizontal axis represents force applied to a bead and the vertical axis, the position of

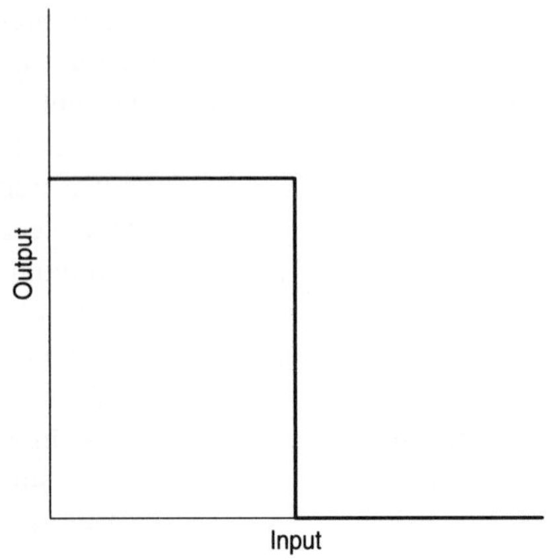

FIGURE 2. The ideal response of a digital inverter.

the bead. In electrical circuits the horizontal axis represents the amplitude of a received bit and the vertical axis, the amplitude transmitted to the next logic stage. The point of this response is that the input may be variable and may have been substantially altered by its passage through the system; nevertheless, a true output is produced. This is absolutely essential in a large system in which signals must be handled many times and would quickly lose their meaning if distortions were allowed to accumulate. Another important aspect of the response in Figure 2 is that there is a large tolerance of variability in the threshold, the value of the input at which the transition between states takes place. As mentioned, such variability is to be expected in mass-produced devices.

The device that produces the response of Figure 2 must have gain. The input signal may well be smaller than the desired output signal because of attenuation during transmission. The highly nonlinear response of Figure 2 is obtained by having a region of very high gain near the center of the device response, a ''threshold'' that separates the region in which the high output is produced from that in which the low output results. The low gain regions nearer to the nominal signal values attenuate unwanted disturbances around the standard digital signal values. In fact, in electrical computing both current and voltage gain are required. Electrical logic circuits must be able to provide fan-out, that is, be able to distribute signals to several receivers. If Figure 2 is taken to represent voltages, sufficient current will be needed to charge the capacitances of a number of receiving devices and the intervening interconnections.

C. LOGIC FOR COMPUTERS

It is possible with this understanding of the function of devices and circuits in digital computers to evaluate the ability of devices to provide the function.

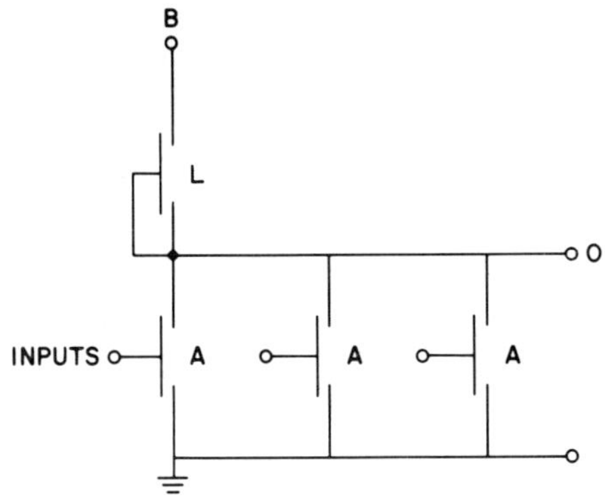

FIGURE 3. A NOR circuit made of field-effect transistors.

The ideal electrical logic characteristic as shown in Figure 2 is exhibited by the electromagnetic relay. A relay establishes a low resistance metal-to-metal contact between an output terminal and a voltage source. Relays can be designed to have high gain; the contacts can carry much more current than that needed to open and close them. Relays were used to build early large computing systems, but relays had a limited future in computers because of their poor reliability, high cost, and slow operation. The search for a better component soon led to the development of computers based on vacuum tubes, which were the basis of the first commercial computers. Tube circuits could operate much more rapidly than relays, but were also limited by high cost and power dissipation and high rate of failure.

The transistor, which first appeared in computers in the late 1950s, made very large machines possible. The unprecedented susceptibility of the transistor to miniaturization led to the integrated circuit and eliminated the cost, power, reliability, and speed barriers of earlier technology, thereby making the construction of machines with millions rather than thousands of components possible. In fact, memories in the computers of today contain tens of billions of transistors.

Functionally, both the vacuum tube and the transistor emulate the relay. A logic NOR circuit that uses an *n*-channel field-effect transistor is shown in Figure 3. Input signals are received by the gates of transistors A. The connections to transistor L cause it to act as a nonlinear load resistor. The application of a positive voltage to the gate of one or more of the transistors A turns the affected transistors on and establishes a connection of the output O to ground, the zero of voltage. If there is no positive voltage among the inputs, the transistors A are all off and the output is connected through L to the power supply B. The ground and the supply B are distributed throughout the system and provide the standard for the digital representation. The input terminals are unaffected by the signal at node O: the calculation flows in one direction.

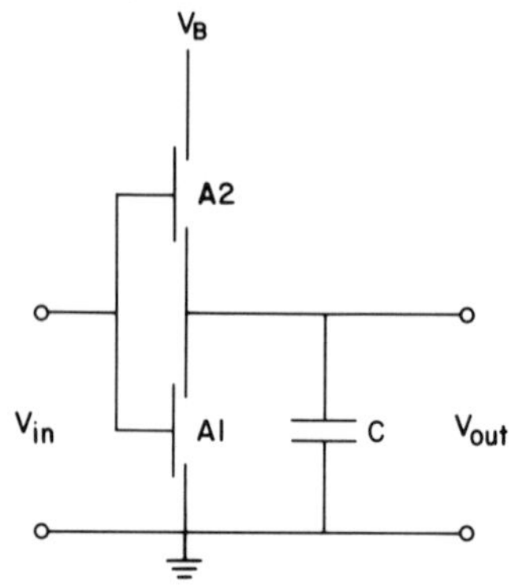

FIGURE 4. A complementary metal oxide semiconductor (CMOS) inverter circuit.

The circuit of Figure 3 is an approximation to the ideal in that when a transistor A is on, a current flows from B through L and A and the finite voltage difference that drives current through A intervenes between the output and ground potential. This small potential that prevents the true standard output from being reached is eliminated in the complementary circuit shown in Figure 4. The circuit is called a complementary metal oxide semiconductor (CMOS), since it is most often implemented with metal oxide semiconductor field-effect transistors. In CMOS circuitry an *n*-channel and a *p*-channel transistor are connected in series between the power supply and ground. An input signal is applied to the gates of both transistors, turning one on and the other off. No current can flow through the circuit in either case. However, when an input signal arrives, a finite time is needed to arrive at a new steady state because of capacitances in the circuit. Let transistor A1 in Figure 4 be *n*-type and transistor A2 be *p*-type. If the output is initially in the high state, the capacitor C is charged to the power supply voltage. When a positive pulse arrives at the input, transistor A1 is turned on and current flows through it, discharging capacitor C. When the output voltage reaches the ground potential, current ceases to flow. If a negative pulse reaches the input, A2 is turned on and current flows until C is charged to the power supply voltage. The output always truly reaches the distributed standard voltages. This feature increases the immunity of CMOS circuitry to the disturbances and component variability found in large systems.

To summarize these examples, transistors, vacuum tubes, and relays are the only three devices that have been used to successfully build large general-purpose computers. The success of all three has been based upon the following characteristics and method of operation, which were emphasized in connection with Figures 3 and 4.

An input signal that may have been attenuated or distorted during transmission produces a large change in the current-carrying capacity of the device, enabling a connection to a voltage standard to be made and broken. The connections allow current to flow until the demands of all receivers for charge have been met, permitting large fan-out. The devices have three terminals, providing excellent isolation of the output terminals from the inputs, so that the flow of signals is in only one direction. Thus, we identify the features that have enabled the great success of electrical logic circuitry and have made modern computing machinery possible:[2-4]

1. The ability to tolerate large variability in the received signal and in the device characteristics
2. High gain; signal smaller than the desired output can change the state of the device and provide fan-out
3. Establishment of the output signal by connection to a standard
4. Three terminals, to allow isolation of input from output
5. Ability to control enough energy to create the electromagnetic fields that effect transmission of signals to other devices

Further, it is known that the operation of inversion is necessary for the realization of a complete set of logic functions. Inversion is easily implemented in transistorized logic by appropriate connections to the standards. Another requirement that must be placed on circuits is the ability to produce signal inversion.

The need for the ability of a circuit to control energy stems from the fact that making a system of many devices work as a unified entity requires communication among the devices. The output terminals of the circuits in Figures 3 and 4 connect through wires to other devices. The passage of electrical signals along these wires involves the creation of electromagnetic fields in regions of space. The energy in these fields must be supplied by current controlled by the driving circuit. For short wires, such as those on chips, the electrical property can be described as a capacitance that must be charged by the signal voltage. The receiving device represents additional capacitance that must be charged.

D. SWITCHES

Switching networks used in communication systems are another kind of large electronic system. A multiplicity of signals enters the switching system and must be routed to diverse destinations. It must be possible to reconfigure the switch to rearrange the interconnections. Reconfiguration is performed by resetting switches that can direct their input to either of two outputs; signals pass through a series of such two-way switches in a large switching network. Complex functions may be demanded, for example, the information arriving at an input port may be multiplexed in time with different time slots intended for different output ports. Large numbers of devices are needed to perform control functions as well as the actual routing operations.

Such large switches share the limitations of logic processors. The many signals in the network simultaneously offer opportunity for crosstalk, and a certain amount of attenuation is inevitable. The problem is the same as that encountered in logic circuits and most of the considerations explained in the previous section apply. Thus, periodic signal restoration is necessary. Restoration is inherently nonlinear and also requires gain. The value of the restored signal is preferably determined by reference to a standard. The necessary operations are most conveniently carried out with transistors.

E. MEMORY

The demands of fast logic processors for information have only been met by electrically addressed random access memory. The demands on the amount of memory and the speed of access to it increase with the size and speed of the associated electronic logic. The memory is supplied by transistorized circuits in contemporary systems. The memories are so large that the provision of a separate wire for access to each memory element is impractical; access is made by matrix addressing. Long lines traverse the memory in orthogonal directions. Information is stored as the presence or absence of a charge on a capacitor in the so-called "dynamic" memories. A switch at each site allows the charge to be transferred to a sensing circuit. The switch is not so different from those used in logic circuits to connect an output to a voltage source and is a transistor for the same reasons. One line of the access circuit turns on a row of transistors, allowing the changes in that row to pass along orthogonal lines to an array of sensors. The devices in the bit read must be able to control a signal large enough to activate a long line and a detecting circuit.

Information is stored in a bistable transistor circuit in the other type of electronic memory. The bistable circuit is essentially a two-stage logic circuit with feedback; two inverters are connected so that the output of each is the input to the other. The requirements on devices are the same as those for logic.

III. CONNECTIONS

Complex logic functions are implemented by a dense network of communication channels. The complex logic enables a system to be programmed to attack a wide variety of tasks, where many tests, branches, and references to memory may be needed. The ability to perform diverse tasks allows the costs of design, development, and hardware to be divided among a large number of applications. The greatest successes have been attained with chips and systems that can serve many purposes, such as microprocessors, memories, gate arrays, and general-purpose computers.

The density of the communication network has had to increase along with the increasing density of components that was made possible by miniaturization. Figures 5 to 7 show the trends that relate to the interconnection of

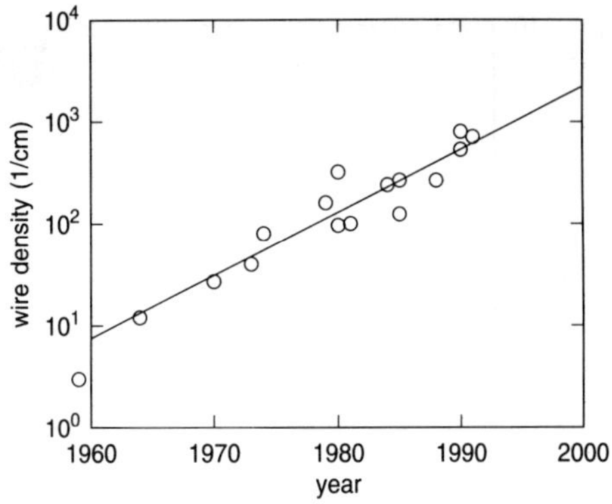

FIGURE 5. The increase in the density of wiring on chip substrates with time.

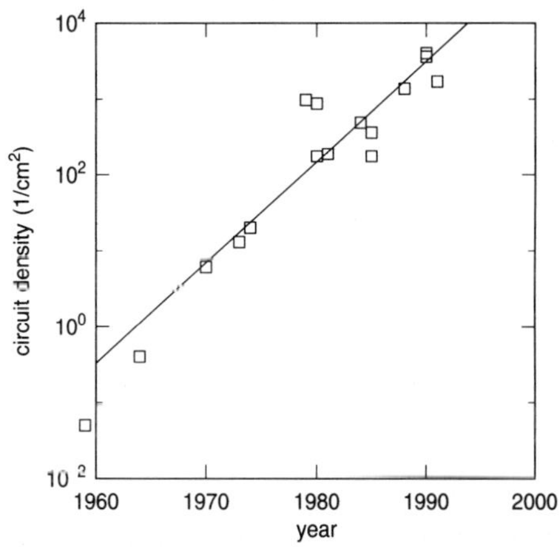

FIGURE 6. The increase in the density of circuits on chip substrates with time.

chips in large computers since the introduction of transistors and furnish a quantitative example of the densification of logic circuitry. The density of components is measured as the number of logic circuits per unit area of a substrate. The density of the communication network is described by the total length per unit area of the channels in which a wire can be placed. A high density of circuits must be supported by a high wire density. The development of these densities since the introduction of transistors into large systems is shown in Figures 5 and 6. There is steady growth of both quantities. Figure 7 relates these observations by plotting wire channel density as a function of circuit density. The line shows the relation $D = 11P^{1/2}$, where D is wire channel density and P is the density of circuits referred to the substrate on which the chips are mounted, called a module, board, or card. Although some

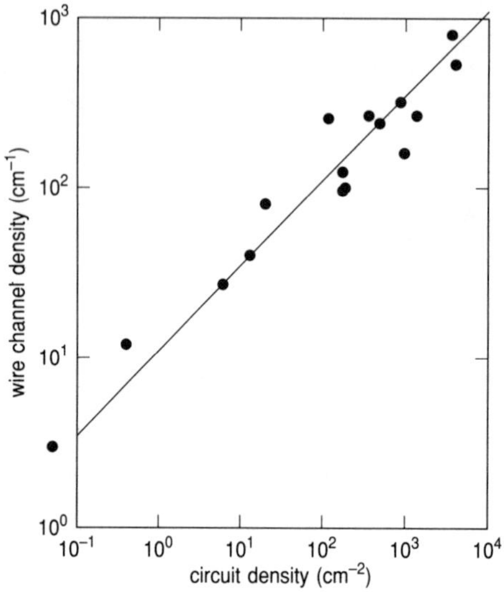

FIGURE 7. The wire density plotted against the chip density for the data of Figures 5 and 6.

of the growth of wire density has been achieved by thinning wires, much of it has come from increasing the number of layers of wire.

The network of interconnections on chips and on chip substrates is designed to implement electronic functions in the collection of components. The most common hardware approach to creating a system is to allow space in the system components in which wires can be placed to form functional entities at a late stage of the manufacturing process. Designing and forming the wires to create a functioning entity is known as "personalization".

The provision of wiring in this way is not solely to allow personalization, which could be achieved in alternative ways. It allows the wiring to be formed at a late stage of the chip fabrication process and therefore permits design changes to be incorporated or errors to be corrected rapidly. These alterations of a design are called "engineering changes", and provision for them is essential, at least in the early stages of constructing a new system. There is sometimes a cost advantage in that different functions can be imbedded in components that have been identical through most of the manufacturing process.

The space allocated to interconnections, that is, the wire channel density, D, must be sufficient to allow a large amount of flexibility in the placement of wires that reproduce a complex logic design. The flexibility may be measured by the number of ways in which the available signal terminals can be connected in pairs.[5] The number of ways in which H terminals can be connected in pairs is $M_0 = H!/2^{H/2}(H/2)!$. Some of these patterns of pairs use a greater length of wire than others. The average wire density necessary to

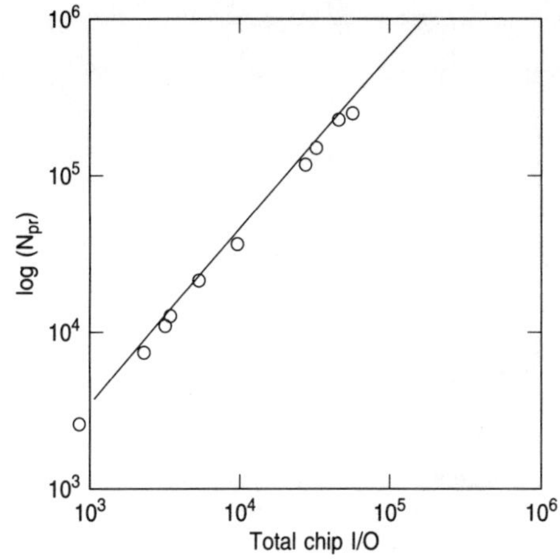

FIGURE 8. The number of ways that a large number of signal terminals can be connected in pairs. The points represent data from multichip modules.

permit the connectivity or versatility M_0 is large and is not needed in practice. Design programs try to select components and wire routes to minimize wire lengths and achieve wire lengths considerably shorter than a random placement would require. The value of M_0 is plotted as a function of H, where H is the total number of chip signal terminals (in Figure 8); the number of ways permitted by the wiring channels on the chip substrates, called multichip modules, in a number of large computing systems is also shown. The deviations of the points from the line are significant, and show that only $^1/_3$ to $^1/_{10}$ of the wire density needed to allow all possible patterns is ordinarily required. Note that the connectivities M_0 are extremely large numbers, $\exp(10^3)$ to $\exp(10^4)$.

The provision of higher and higher wire densities has required continuing innovation in the form of both thinner wire channels and more layers of wire. The difficulty of fabricating the wiring network increases with both the density of the wire channels and the area that must be wired at that density. The finer the wire, the smaller the defect that can cause an open or short circuit; and the larger the area, the greater the probability that it contains a defect. The author has proposed that the area $\times$ $(\text{density})^2$ product is a measure of progress in wiring and found that technology has enabled the product to increase at a rate greater than one order of magnitude every 5 years.[5] The numbers and examples of data are presented in Figure 9.

IV. NOVEL CONCEPTS

In spite of the astounding success of the above approach to automated information processing, there is no shortage of proposals for performing logic

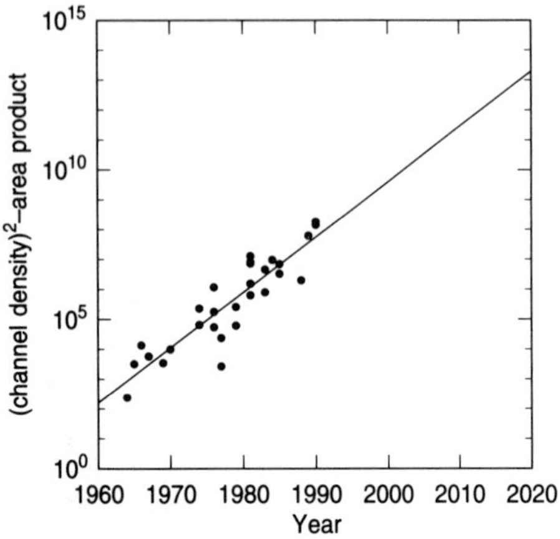

FIGURE 9. The product (area × square of wire density) for various chips and multichip modules as a function of time.

with devices that do not possess the essential attributes that have been identified. An examination of some of these illustrates and reinforces the fundamental nature of those requirements.

A. BISTABILITY

Bistable circuits based on various devices that have a negative resistance region have received much attention. An example that attracted substantial support in the 1960s is the tunnel diode. The current-voltage characteristic of a tunnel diode and the circuit that was proposed to use it for logic are shown in Figure 10.[6] The stable states of the circuit are determined from the load line in the figure, and are designated A and B. The idea of the circuit is that it is initially set to state A. If current injected through one or more of the resistors R is sufficient to cause the total current to exceed the peak current at P, then the circuit will switch to state B and the superfluous current will leave through output terminal O to be input to further logic stages. If the initial states and inputs are such that one input through R will cause the circuit to switch, an OR function is implemented; if all inputs are needed to cause switching, the operation represents an AND.

Substantial efforts were devoted to attempts to secure a place for tunnel diodes in digital logic. It is worth examining the reasons for their failure. As can be seen from Figure 10, a high gain can be obtained by biasing the circuit so that state A is very close to the peak of the current-voltage characteristic. This means, however, that there will be a very large sensitivity of the switching threshold current to the height of the peak, which will vary for the reasons already described. In fact, if the height is not well controlled, the intersection A will not exist in some diodes. Gain depends on the ability of technology

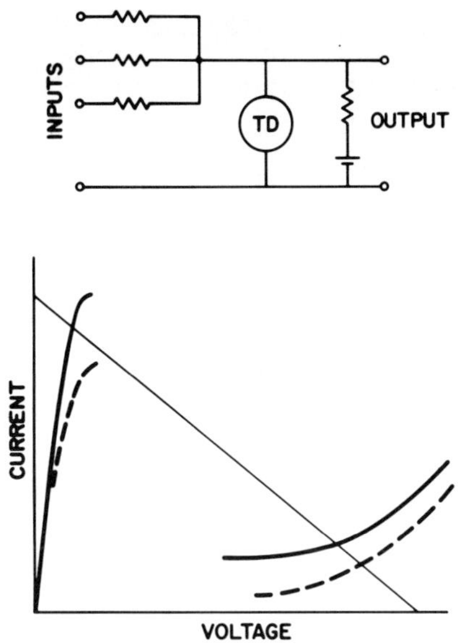

FIGURE 10. The tunnel diode circuit and its current-voltage characteristic.

to control the height of the peak and the value of the resistor that determines the load line; there is little tolerance of component variability. In practice the i-V curve could not be reproduced sufficiently well to allow working tunnel diode logic circuits to be consistently manufactured.

Other problems also existed with the circuit. The output signal is not derived from a standard; it depends on the device. This point and the preceding one are illustrated by the dotted characteristic in Figure 10, a modification of that given by the solid line. The variability in device characteristics is reflected in variability of the current carrying information to the following stage. There is poor isolation of input from output; switching will send a signal back through the input resistors R. The response of a diode will depend to some extent on the status of the following stage rather than being determined by its inputs alone.

There is no inversion; current in causes current out. In addition, a separate resetting operation, such as turning off the power, is needed to prepare the circuit for another logic operation.

The circuit considered is also subject to "critical slowing".[7] Gain requires that the load line pass close to the peak, as shown in the figure, so that during switching there will be a slow passage of the diode voltage past the peak; there most of the current will flow through the diode rather than being available to the output, where the capacitance of wires and following stages must be charged to complete the operation.

Negative resistance is a common and easily achievable phenomenon in electrical devices, and suggestions for use of it to perform logic operations

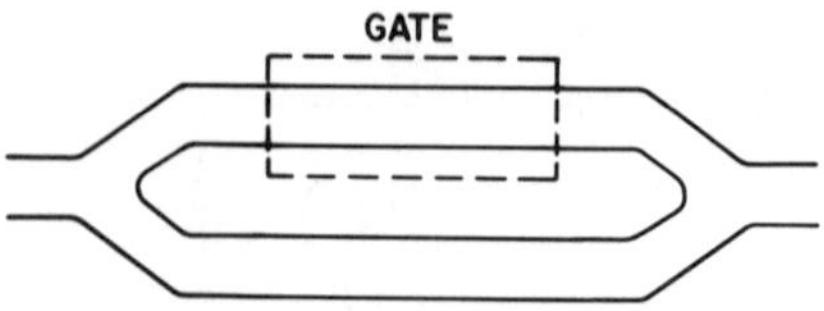

FIGURE 11. A quantum interference device.[8]

abound. These are essentially the same as the tunnel diode proposal just described, and suffer from the same basic high sensitivity to variability of component parameters and the dependence of the output signal on the device characteristics. In particular, the same difficulties are encountered in attempts to perform logic with optical bistability, which has also received attention recently. All of these negative resistance-bistability schemes can be demonstrated with a few stages in a laboratory, where favorable devices can be selected and voltages and other circuit parameters carefully adjusted to achieve operation. Such detailed attention to individual details is not possible when systems of thousands or millions of components are used and where reproducible and reliable operation over long periods of time is essential.

B. QUANTUM INTERFERENCE

Devices based on quantum interference in solids have received considerable attention recently.[8-10] The concept is shown in Figure 11. A device has two parallel channels through which current may flow. It must be so small that electrons can pass through the channels coherently, that is, without the phase of the electron wave being destroyed by scattering between the ends of channels X-X and Y-Y. Then if the channels are identical, the waves will interfere constructively at the output end and a current will flow. If a magnetic field is applied so that magnetic flux passes through the region between channels X-X and Y-Y, then the transmittivity of the structure changes in accord with the well-known Aharanov-Bohm effect. More in keeping with the methods of electrical logic, a similar difference in phase between the electrons exiting the channels arises if there is a difference of electric potential between them. The difference in electrical potential can be produced by an electric field in the *y* direction supplied by some control electrode. The difference could also be obtained in a Mach-Zehnder interferometer configuration, with an electrode like the gate of a field-effect transistor that is placed on only one branch and changes the energy of the states in that branch.[8-10]

This kind of device would have to be used at cryogenic temperatures to prevent phonon scattering from destroying the phase-coherent propagation of the electrons. The potential difference between the two channels could not be too high or too low; it would have to be just right to cause out-of-phase interference. There would be little immunity to the effects of signal variability caused by attenuation and noise. Since the difference in phase at the end of

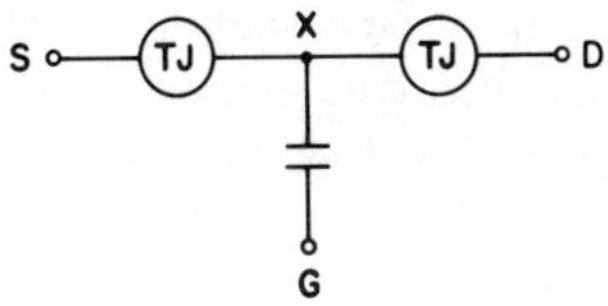

FIGURE 12. A circuit that shows the Coulomb blockade effect.[10]

the channels depends on their length, close control of the lengths would be essential; high precision would be demanded of the fabrication process. Any differences between the channels that might arise from other sources, such as strain or a temperature gradient or surface scattering, would also affect electron propagation and operation of the device. All of these difficulties would plague attempts to build a large system from such devices.

C. COULOMB BLOCKADE

Single-electron devices have excited recent interest. The circuit configuration of a recently reported lithographically fabricated structure that exhibits single-electron effects is presented in Figure 12.[11] Two tunnel junctions TJ of small capacitance are connected in series. The voltage across each of the tunnel junctions depends on the charge on the conductors that connect them, the node X. A change in the charge by one electron, charge q, changes the voltages by an amount of order q/C_J, where C_J is the capacitance of a tunnel junction. Tunneling of an electron through one of the junctions changes the potential of node X, which changes the tunneling probabilities. If C_J is as small as 0.1 fF, then voltage steps of around 1 mV appear in the current-voltage characteristic of the circuit S-D as the charge on the node changes by one electron. A potential applied to the terminal G in Figure 12 also changes the charge on node X and the position of the voltage steps. The small junctions required to produce such small capacitances were realized by Fulton and Dolan[11], and the predicted effects in the current were observed. The effect is named "Coulomb blockade" because of its dependence on the electrostatic repulsion of elementary charges. It is overwhelmed by thermal energy unless the temperature is so low that $kT \ll (q^2/C_J)$.

The effects become more complicated in situations in which the energy levels within parts of the circuit of Figure 12 must be viewed as discrete; that is, they are separated by an amount comparable to the Coulomb charging energy.[12,13] Such conditions are easily produced in III-V semiconductors because the low effective masses of electrons in many of the compounds cause large separation of energy levels in narrow wells. In addition to the potential differences, tunneling depends on coincidences of the energy levels in the device elements. The electric current then depends on the interrelations between three energies: the Coulomb energy, the quantization energy, and the thermal energy. The complex phenomena that can occur have provided a fertile field for both experimental and theoretical investigation and have spawned a large and rapidly growing literature that will not be pursued further here.

The idea of the Coulomb blockade has been long known and explored. The key to the clear demonstration of the effects in Reference 10 is the isolation of node *X* between two tunnel junctions. The single-electron tunneling configuration shown in Figure 12 is equivalent to a three-terminal device. Its potential for use in applications is severely limited, however. The effect produces a rather minor perturbation of the tunneling current-voltage relation in comparison with the on-off response that can be obtained with a transistor. The very small capacitances needed to observe the Coulomb blockade are difficult to fabricate and are easily overwhelmed by stray capacitance. The energies involved are so small that they are dominated by thermal energy except at cryogenic temperatures. The energy levels are broadened and the Coulomb blockade effect is obscured unless the residence times of electrons on the node are long. It is required that the time, τ, satisfy $\tau \gg 2C_J h/q^2$, implying that impedances are much larger than $h/q^2 = 25000\ \Omega$. Only small currents can be controlled because of the small amounts of charge involved and the high impedances. As a result, communication with other devices is very restricted; there is no ability to drive lengthy lines or to provide large fan-out. Although the physics is revealing, the potential for application to electronics is small.

D. CELLULAR AUTOMATONS

High gain does not come easily. In fact, both the transistor and its predecessors, the vacuum tube, depend on the same mechanism for gain, namely, the ability of an immobile electric charge to control a highly mobile electric charge through Coulomb attraction. In view of the dominance of this unique principle of amplification for eight decades, the difficulty of finding another kind of device with the gain and current-carrying capacity that enables it to provide fan-out and feed current to long interconnections is not surprising. Thus, there is strong temptation to propose that the many lengthy connections that characterize the logic in computers can be eliminated. After all, it can be argued that cellular automatons can emulate a Turing machine and that a Turing machine can do anything another computer can do. If this were a realistic way to approach computing, however, one would not have to wait for the invention of a suitable new device; cellular automatons could be built with transistors, and there are undoubtedly functions for which something like a cellular automaton is ideally suited. (After all, if one wishes to explore the game of LIFE, why not build a machine with an array of simple logic functions and the appropriate nearest neighbor connections at each site?) However, most investment in computing facilities is made with the intent that it will provide the means to handle a variety of problems. Tasks frequently require references to distant memory locations; tests can indicate that a branch to a different sequence of instructions is in order or that a calculation has been completed. It has never been shown or argued that such functions can be realistically implemented without complexity in the interconnection pattern. Many simulations show that cellular automatons can be devised to do a particular task but do not reveal a route to diversity; it is hard to imagine a way to program a cellular automaton to perform a variety of functions. In fact, it is much simpler to simulate the cellular automaton on a general purpose computer, rather than vice versa.

Cellular automatons in which each cell is a complex information processor in its own right can and have been built. These systems are not relevant to the inability of devices to support communication because the considerations that apply to computers then apply within each cell.

Systems with nearest-neighbor interactions can function as shift registers, however; and these have found application in technology. One example will be discussed in the following section. Another is the charge-coupled device (CCD).

E. MAGNETIC BUBBLES

History, in fact, provides an example of a logic proposal that is restricted to near-neighbor interactions, logic with magnetic bubbles.[14,15] Here magnetic bubbles rather than electrons are moved from site to site. A magnetic bubble is a cylindrical domain in a thin film of an anisotropic ferromagnetic material.[16,17] Patterns of soft ferromagnetic material on the surface of the film influence the energetics of the bubbles and create preferred positions for them. An externally applied magnetic field can cause the bubbles to move between adjacent preferred positions of a properly designed ferromagnetic overlay.

The main interest in magnetic bubble technology is motivated by its use as memory. A bubble memory consists of a large lithographically prepared array of sites. The presence or absence of a bubble at a site carries information. The memory operates as a clocked shift register, with the bubble pattern being moved from site to site by a periodically varying magnetic field. A writing station enters information into the shift register and a reading station detects it as it arrives. The form of a bubble is determined by magnetostatics. Its form is, therefore, restored at each site; energy, if necessary, is drawn from the driving field to do so. Good uniformity of the magnetic film is required. Magnetic bubble technology has reached a high state of development, multimegabit memory chips have been made, and bubble memory systems are available commercially. The simple structure needed to define a site has permitted high levels of integration and low cost memories to be achieved. Bubble circuits are attractive in that they are nonvolatile; the bubble configuration persists without a source of power. However, shift register operation and a low velocity of motion cause long access times. Bubble memories have only captured a small niche of the memory market in the face of competition from much faster semiconductor memories.

A repulsive interaction between bubbles permitted logical functions to be implemented. Elementary logic functions such as AND and OR have been demonstrated.[18] Bubbles can be replicated to provide fan-out, with a cost in space and time. Fan-in is limited by the short range of bubble interactions by geometry; only two or three bubbles can be close enough to a site to influence it. Bubbles have a limited range of stability,[17-19] leading to concerns about operating margins. However, long times are involved in the shifting of information through a distance, and there are no reports of logic arrays of any size.

F. ARCHITECTURES

It is also tempting to embrace the thought that the deficiencies of a device will be compensated by the development of some novel architecture that will not suffer from lack of desirable attributes of the device. History does not encourage such optimism. Those who have waited for someone to invent an architecture to utilize their device have been disappointed. Developments in the architecture of electronic computers through the era of their existence have involved ways of using more devices to achieve greater system speed by doing more things in parallel. None of these have offered opportunities for avoiding the need for signal standardization, high gain, noise immunity, and ability to communicate that might allow new device concepts to make an inroad into the reliance on transistors. Although continuous innovation has marked computer technology, introduction of basically new devices has only occurred outside of the logical circuitry. For example, memory and storage technologies have evolved through cathode ray tubes, delay lines, cores, and disks. Light emitting devices have been introduced for optical fiber communication between separated parts of a system and for optical disk storage.

V. MINIATURIZATION AND MOLECULES

A. MOLECULES AND ELECTRONICS

A number of fairly recent events excite interest in the association of molecules with electronics and information processing. One is simply the approach of the dimensions of solid-state electronic structures to those of molecules. The long continuation of the miniaturization of electronics allows the size of electronic components to be extrapolated to the size of molecules. The changes with time are shown in Figure 13, which continues the plot of Figure 1 through the coming decades. The areas plotted in Figure 1 have been replaced with estimates of the number of atoms in the physical volume of the entity that stores a bit in the extrapolation. There will be a large gap between the sizes of molecules and those of bits for a long time to come, but the narrowing of the gap has been surprisingly durable and will continue. Another facet of the convergence of sizes is that films with thicknesses less than 100 Å, less than about 30 atomic diameters, are used in electronics.

New tools that enable investigation and control of phenomena on a molecular scale also stimulate interest in molecules. The scanning tunneling

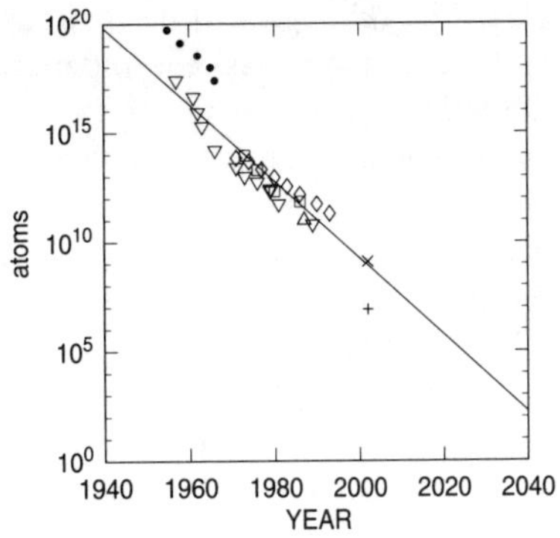

FIGURE 13. Estimates of the number of atoms used to store a bit of information in various technologies as a function ftime. The symbols have the same meaning as in Figure 1. Points have been added. The dots represent magnetic cores. The + and the × are experimental memories that have been rather arbitrarily placed 10 years in the future.

microscope is the outstanding example and resolves detail on an atomic scale. It has also been used to move atoms one by one to create simple structures.[20,21] Imagination is easily extended to the possibility of building molecules with such a tool. The new techniques of micromachining, the use of the methods of microelectronic fabrication to form miniature mechanical structures, may also converge with molecular science. Opportunities to coin new words with the prefix "nano" abound as a result of these advances.

Still another tool is the laser, with intense narrow lines that lend a new dimension to molecular spectroscopy. The great spectral resolution of laser light has suggested possibilities for storing information in a single molecule.[22,23] The molecules in question are made of about 40 atoms. Matrix addressing of three-dimensional molecular storage systems with perpendicularly directed light beams is proposed.

The growing development of electronic sensors is an additional point of contact. Many sensors depend on the ability of molecules that are attached in some way to a field-effect transistor (FET) to affect the electrical characteristics of the transistor. Much effort is devoted to achieving specificity, response to one molecule to the exclusion of others. The small size of FET sensors lends them high value as biological sensors.

The discovery of organic semiconductors, superconductors, and ferromagnets, long considered the domain of metals and semiconductors, has an obvious relation to the needs of electronic technology. Large optical nonlinearities in organic materials may have an important role in optoelectronics. In fact, the great versatility of organic molecules and the existence of techniques for creating new ones with predictable properties may form a base for applications of molecular crystals and molecules in technology.

Increased understanding of the molecular basis of biological processes has led to the realization that many reactions can be viewed as information processing by molecules.[24,25] Although still not thoroughly understood, it is becoming clear that molecular shape plays an important role in biochemical events. The recognition of shape by molecules is a complicated form of information processing. It appears, then, that a change of shape caused by some external stimulus should affect biochemical reactions. Such effects may have more significance for sensing than computing and other information handling. Will micromachined parts reach a stage that permits them to participate in molecular shape recognition?

The transduction of light for purposes of vision and photosynthesis hints at a relation to molecular electronics. The excitation of molecules to excited states by light has long been familiar and plays an essential role in vision and photosynthesis. A single photon can deliver enough energy to a molecule to change its electronic state, and the dependence of the physical configuration of a molecule on its electronic state is also well known. Such effects can be quite striking in complex molecules; witness the well-studied photo-induced isometry in rhodopsin.[26] Photosynthesis depends on series of transfers of electronic energy produced by photoexcitation.[27] Although the energy efficiency is low, only about one third of the energy received from photons is stored as chemical energy; it rivals the best diode solar cells.

Therefore, a rather broad collection of topics that may be called molecular electronics exists. An obvious question is can molecular devices play a role in large information-handling systems? Can a molecule reproduce the functions that components in systems perform? The enormous variety of molecules and a confidence that a vast universe of them remains to be discovered lend optimism to this line of thought. All of these observations stimulate interest in molecules as electronic components.

It is appropriate to ask what one can say about molecules in the light of our understanding of what computational devices must do. The great success of the transistor in electronics sets it apart as a model of what other electrical devices should be like. In the first place, a molecule that is to emulate an electronic device must be susceptible to some sort of external influence. It must be possible to change the state of the molecule in some significant way by application of an appropriate signal.

Powerful supercomputers are another technological advance impacting molecular studies. Numerical studies of molecular wave functions lead to predictions of molecular properties and molecular transitions.[28,29] A molecular transition that changed a molecule to a state permitting electrons to pass through the molecule from one in which that cannot happen would seem to reproduce the action of a relay if the change can be produced by application of an electrical signal to a third terminal. The suggestion that such a transition, caused by exposing a molecule to a strong electric field, may exist is therefore most interesting and relevant.[29] Energetics would seem to make such an effect uncommon. One would want to have an energy barrier of at least 10 kT between states for stabilization against thermal transitions. Few materials will endure an electric field of 10^6 V/cm, and this high electric field would have to act through 25 Å to supply 0.25 eV to a device at 300 K. A large separation from thermal excitations is part of the reason that semiconductors with energy gaps of 1 eV or more are preferred for electronic applications. The electronic excitations in vision and photosynthesis are induced by 2 eV photons, even further from thermal interference.

However, as often emphasized above, evidence that something one can measure has occurred does not make a logic device; it is necessary to look at other attributes and to the quantitative aspects of an effect to see whether it is a device. One must examine the other aspects that determine suitability as an electronic device. Is there gain in the sense that the control current is smaller than that which is controlled? Can the change in molecular structure be produced by a range of control signals or is only a narrow range of input signals acceptable? In other words, how successful is the effect in emulating the relay? Is the state of the molecule sensitive to the parameters of the output or is there good isolation of input from output? If the answer to such questions is affirmative, then it is worthwhile to examine questions of speed; stability; contacts; temperature sensitivity; and many others, for example, issues of manufacturability and assembly. Unfortunately, it appears that little thought or effort has been applied to characterizing molecular events in terms of parameters that would be useful in evaluating their suitability for application as logic devices.

One problem might be relieved by the use of molecules as electronic devices. Reference has been made to the imprecision of manufactured components. The nature of a molecule is set by the quantum physics of chemical bonding: presumably all molecules of a given composition are the same. As in the case of manufactured electronics, external influences can modify properties, however. Strains from attachment to a substrate will have an effect; inhomogeneities in a substrate surface may also modify properties. Gradients of temperature will create differences between different regions. Also, one must ask what is necessary to ensure that all of a large number of molecules actually do have the same chemical composition. Quantitative questions will arise. What standards of reagent purity will be required in order to obtain satisfactory yield in an assembly of 10^6, 10^8, or 10^{10} molecules? What sort of clean environment will be wanted to control the contamination that plagues electronic manufacturing?

Hopes are sometimes offered that a self-assembling molecular system will evolve. This would certainly solve many problems. Again, however, one must look beyond the molecules themselves to evaluate the usefulness of this concept. As we have argued, communication between nearest neighbors creates a system of limited value. How can self-assembly reproduce the long dendrites and axons of neurons? One may also find hints of potential problems in crystal growth, which is somewhat analogous to, though simpler than, molecular self-assembly. The perfection of crystals is limited by native defects (vacancies, interstitials), dislocations nucleated at a substrate or seed, and spontaneous nucleation of unwanted crystallites. How prominent will similar things be in molecular self assembly?

B. MOLECULES IN SYSTEMS

The prospect of an electronic component being constructed of a number of atoms that might also form a large molecule naturally inspires the thought that the same atoms could be used to build a molecule instead of a conventional electronic component. Let us assume molecules that can replicate the transistor function have been found and look briefly at a few quantitative problems. The use of molecules would enable a giant step in the direction of miniaturization and high component density. It is to be expected that use of these molecules as a system will require that a large number of them be placed on a substrate on which they can interact with one another. A great advance over conventional electronics must be sought, and a transistorized logic chip will be used as a guide to what is wanted. Now we try to imagine the physical form of an array of molecules that can perform logical functions electrically as the transistor does. As this will no doubt be far in the future, an extrapolation of semiconductor circuitry will give some idea of what the molecular machine must surpass. As this molecular technology will not be available for decades, the model chip is a projection of the characteristics of transistorized chips into a distant future, when its validity is speculative and an alternative to the projected chip may be wanted. Development activities aimed at over 10^8 transistors on 64 and 256 Mb DRAM memory chips are already underway and extension to 10^9 is in the near future.[30] Thus, the integration of 10^8 logic circuits on a chip is taken as a minimal goal.

The nature of a 10^9 transistor logic chip can be estimated by straightforward extrapolation of three decades of the history of that technology. First, it can be guessed that this level of integration will not be attained for 30 years, which allows ample time for molecular electronics to reach an advanced state of development. The extrapolation suggests that the logic chip 30 years hence will contain 10^8 logic gates and have an area of 4 cm^2, or a space of 2 μm by 2 μm per gate. In spite of continued miniaturization, this is still far larger than molecular dimensions. Imagine, therefore, that the 10^9 molecules are placed on a 1000 Å grid and used to build 10^8 logic circuits. Further, let the 10^8 circuits be organized into 100 processors similar to contemporary microprocessors, each containing 10^6 circuits. Then compare these 10^6 circuit assemblies with an electronic chip.

The hardest part of making a logic chip is making the wires. The amount of wire, measured in the right way, grows faster than the number of circuits. The number of wire channels that must be provided in each logic cell has been estimated.[31] Extrapolating the results of Heller et al.[31] to 10^6 gates suggests a need for about 60 wire channels passing through each logic cell. If all of these channels were placed side by side each would have a width of only 50 Å. Wire channels are not provided in this way; however, the demand for channels is always so great that several layers of wire are used to supply it. Two layers are needed in any case to allow wires to run in x and y directions. If there are K layers, then the width of each channel can be K × 50 Å. At present four layers of wire are formed on some chips, but over 30 can be formed in multichip packaging substrates. Assuming that the methods of packaging technology have been adapted to make 12 wire layers possible on chips allows wire channels to be 600 Å wide, providing a wire density of 2×10^6 cm/cm^2. If half of these channels are actually occupied by a wire, 1 km of such wire would be needed in this system that has an area of 0.1 cm^2 — a difficult, but perhaps not impossible, prospect.

The feasibility of the projected wire can be viewed in the light of past achievements. The difficulty of fabricating high wire densities increases with the density and also with the area to be wired; the larger the area, the more likely it is to include a defect. The author has suggested that the state of electronic wiring technology may be measured by the product of the square of the wire density and the area. This product is plotted in Figure 9 for various chips and multichip substrates as a function of time, and a line fitted to the points is extrapolated to future years. The value of the product for our hypothetical molecular chip is $(0.1 \text{ cm}^2) \times (2 \times 10^6/\text{cm})^2 = 4 \times 10^{11}$. The extrapolation predicts that this value will be achievable in 20 years.

The wires in the 600 Å channels would have a resistance of about 30 Ω/μm. The average wire would be 7 μm long and have a resistance of 200 Ω. This high value would be a source of difficulty; and, although these numbers cannot be taken literally as future reality, they show that providing sufficiently low resistance wire is a problem that is likely to be severe in the futuristic speculative molecular technology.

Although it is unlikely that this extrapolated chip will ever be built, and the projected schedule may be far off the mark, it is an example of the kind of complexity that might be expected of electronics in the early 21st century.

One of the important limitations on the performance of chips is the ability to remove heat from them. Extrapolated cooling technology will be able to remove 2500 W/cm^2 from such a chip. While this number is hardly believable, 30 years ago the same could have been said of current technology. On the other hand, skepticism about the capacity of technology to handle 2500 W/cm^2 is reinforced by noting that the power density at the surface of the sun is 7500 W/cm^2! We regard 250 W/cm^2 as closer to a realistic limit.

Now consider heat production. A minimum value for the energy dissipated in an operation can be obtained as the energy of charging and discharging the wires, CV^2, where C is the capacitance of the wire associated with a logic gate and V is the signal voltage on the wires. The signal swing must dominate the thermal voltage, and its minimum value is several tenths of a volt. The capacitance per unit length of wire has not changed much with time; it differs from the permittivity of the vacuum only through a dielectric constant and a shape factor, and is taken to be C = 1 pF/cm. Then, if half of the 60 channels are occupied by wire, each circuit must drive 10 μm of wire on the chip, dissipating 0.25 fJ of energy per operation. The 250 W/cm^2 cooling density allows 25 W to be dissipated in the 100 processors, or 0.25 μW per circuit. The circuits can then be operated once each nanosecond.

Finally, let us examine the currents required in these circuits. The charge that must be placed on the wires is about 0.5 pC/cm. The 10 μm of wire uses 0.5 fC of charge each nanosecond, an average current of 0.5 μA.

It appears the factors that threaten to limit the miniaturization of conventional electronics will be encountered much more abruptly in a transition to molecular circuitry. The fact that the pursuit of miniaturization causes power dissipation to become an increasingly difficult aspect of technology is very fundamental; the number of components increases inversely as the square of dimension, while capacitance decreases only linearly with dimension. The capacitance per unit area that is charged from the power supply in each cycle of operation increases. The increasing wire resistance also has a fundamental origin. Simple scaling predicts that the cross-sectional area of a wire decreases as the square of the scaling factor, while the length decreases linearly. This effect has been partially countered by increases in the number of layers of wire on chips and chip substrates. However, it is difficult to accept the idea of an unlimited number of wire layers, and it is hard to escape the reality of a resistance limit.

Of course, a whole new technology would be needed to duplicate electronic solid-state circuitry with molecules. Prominent among these would be electrical contacts and attachment. Making low resistance contacts to devices and ensuring adhesion between different materials are persistent obstacles that must be overcome even in the development of more conventional circuitry.

C. MOLECULES IN CHEMISTRY

The important results of molecular events belong in the realm of chemistry. Photosynthesis results in the storage of energy in a glucoselike molecule through the reaction:

$$6CO_2 + 12H_2O + \text{photons} \rightarrow 6O_2 + C_6H_{12}O_6 + 6H_2O$$

The photo-induced isomerization in rhodopsin is quickly turned into a change in chemical bonding. Presumably an electrically induced isomerization of a molecule could have similar consequences. Indeed, changing chemistry through the application of an electric field could impact both fundamental and practical science and would seem a more fertile field than a quest for a molecular transistor.

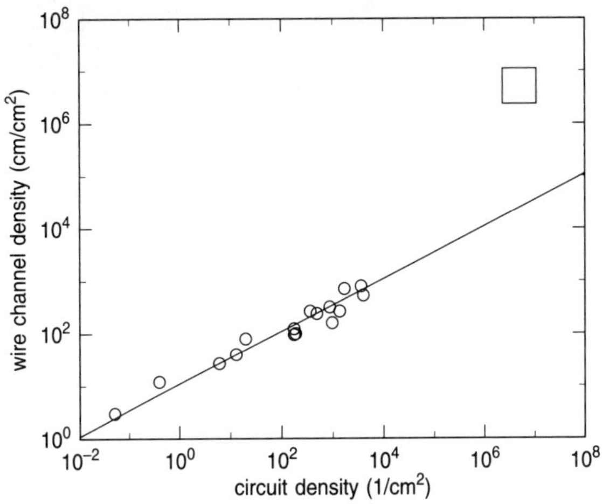

FIGURE 14. Extrapolation of the density of wire on logic chips compared with that of the cerebral cortex.

VI. THE BRAIN

The comparison of artificial electronic systems with the brain has long attracted interest,[32] in spite of deep-seated differences. The quantitative facts concerning the brain are perhaps even less firmly known than those of a chip containing logic circuits 20 or so years from now. Nevertheless, here the cerebral cortex is compared with a chip. The area of the average human cortex is about 1500 cm^2. It contains something like 10^{10} neurons, each of which consists of a central body and efferent and afferent fibers. Connections to other neurons are made via synapses on the long fibers.

There is great uncertainty regarding what in the cortex should be regarded as equivalent to an elementary logic circuit. A neuron is an obvious choice, but may have fan-in and fan-out of 1000 or more; and its function is probably much more complex. The analogy of the long fibers to electrical wires seems more obvious, but may not be completely correct. Pursuing the analogies in spite of such doubts leads to a view of the cortex as having a density of 5×10^6 logic gates per square centimeter. A very rough guess is that the cortex contains 10^5 km of fibers, an interconnection density of 5×10^6 cm/cm^2. Figure 14 compares these numbers with electronic systems (Figure 7) and suggests that a chip of the size of the cortex is much richer in interconnections than a chip with 10^{10} logic gates would be. Although Figure 3 was based on multichip substrates, data for chips would show only a factor of two to four higher wire density.

VII. MEMORY AND STORAGE

The highest development of miniaturization and close packing of components is found in memory. The seemingly unlimited ability of information handling systems to utilize enormous amounts of memory has stimulated massive efforts to minimize costs per bit by maximizing the number of bits on a chip. The same force drives a similar trend in storage devices in which information is stored on an unstructured surface and accessed by motion of the surface and/or a reading system. Advances in storage (mainly magnetic disk files) were included in Figure 1, in which the history of the area of a surface required to store a bit of information in various technologies was recounted. When it is recalled that the dimensions of the surfaces involved range from 1 to something like 30 cm and that a great many such surfaces are incorporated in a single system, the astronomical nature of the numbers involved may be appreciated.

The storage of information in the state of a molecule is an intuitively attractive concept. At least at first sight, a less complicated function is required of a molecule that represents a bit of information than of an active electronic device; some means of setting the state of the molecule is needed; and the state must be retained for an indefinitely long time. A property that enables the state of the molecule to be determined, optimally without affecting the state, is also required. Optimally, too, either state could be changed to the other with comparable amounts of effort; the memory would be "read-write". The proposals of Parthenopoulus and Rentzepis[22] and Basche and Moerner[23] presumably satisfy these requirements.

Molecules could also, however, be mounted on a disk and accessed by mechanical motion. Ordinary disk storage is valued because of its low cost per bit, which is attainable because a manufactured structure is not used for each bit. The mechanical motion makes access to the bits slow, first because velocities are low (thousands of centimeters per second) in comparison with the velocity of electromagnetic waves; additionally because access is partly sequential, a track must be followed until the desired information is reached.

Alternatively, if molecules with the proper electronic interaction were found, implementation of a shift register memory can be imagined.

VIII. CONCLUSIONS

Modern artificial information processing is accomplished with systems containing a great many devices. Certain requirements that must be placed on the devices from which these systems are built can be identified from the history of logic. The differences between the environments in a large system and in a laboratory are often not recognized, with the result that the essential attributes are missing in device proposals. Thus, although many proposals for devices have been put forward, only three — the relay, the vacuum tube, and the transistor — have proven able to meet the requirements and form the basis of large computing systems.

Two themes have been emphasized. Tolerance of variability in signals and in component parameters is essential. Great flexibility in placing interconnections plays a vital role. Although logic operations with devices that are unable to send signals to a distant part of the system have been demonstrated, history shows that none have been able to evolve from the laboratory to the logic marketplace.

A variety of reasons for associating molecules with electronics can be advanced. The existence of molecules that can mimic the behavior of a relay has been suggested, but remains to be demonstrated. Molecules that can replace transistors in electric circuits would be much smaller than transistors and offer an opportunity for a giant step in the direction of higher component density. They would, however, also hasten the impending collision between advancing miniaturization and fundamental issues of resistance and power dissipation.

REFERENCES

1. **Williams, M. R.,** *A History of Computing Technology*, Prentice-Hall, Englewood Cliffs, NJ, 1985.
2. **Keyes, R. W.,** *The Physics of VLSI Systems*, Addison-Wesley, Wokingham, 1987.
3. **Keyes, R. W.,** Physics of digital devices, *Rev. Mod. Phys.*, 61, 279, 1989.
4. **Keyes, R. W.,** Limits and challenges in electronics, *Contemp. Phys.*, 32, 403, 1991.
5. **Keyes, R. W.,** The power of connections, *IEEE Ckts. Dev.*, 7(3), 26, 1991.
6. **Gentile, S. P.,** *Theory and Application of Tunnel Diodes*, Van Nostrand, Princeton, NJ, 1962, chap. 8.III.
7. **Garmire, E., Marburger, J. H., Allen, S. D., and Winful, H. G.,** Transient response of hybrid bistable optical devices, *Appl. Phys. Lett.*, 34, 374, 1979.
8. **Datta, S., Melloch, M. R., Banyopadhyay, S., and Lundstrom, M. S.,** Proposed structure for large quantum interference effects, *Appl. Phys. Lett.*, 48, 487, 1986.
9. **Capasso, F. and Datta, S.,** Quantum electron devices, *Phys. Today,* 43(2), 74, 1990.
10. **Fowler, A. B.,** U.S. Patent 4,550,330, 1985.
11. **Fulton, T. A. and Dolan, D. J.,** Observation of single-electron charging effects, *Phys. Rev. Lett.*, 59, 109, 1987.
12. **Beenakker, C. W. J. and H. Van Houten, H.,** Quantum transport in semiconductor nanostructures, *Solid State Phys.*, 44, 1, 1990.
13. **Averin, D. V. and Likharev, K. K.,** Single electronics: correlated transfer of single electrons in systems of small tunnel junctions, in *Mesoscopic Phenomena in Solids*, Altshuler, B. L., Lee, P. A., and Webb, R. A., Eds., Elsevier, Amsterdam, 1991, chap. 6.
14. **Minnick, R. C.,** A system of magnetic bubble logic, *IEEE Trans. Comput.*, C-24, 217, 1975.
15. **Lee, S. Y. and Chang, H.,** Magnetic bubble logic, *IEEE Trans. Magn.*, MAG-10, 1059, 1974.

16. **Bobeck, A. H., Fisher, R. F., Perneski, A. J., Remeika, J. P., and van Uitert, L. G.,** Application of orthoferrites to domain-wall devices, *IEEE Trans. Magn.*, MAG-5, 44, 1969.
17. **Gianola, U. F., Smith, D. H., Thiele, A. A., and van Uitert, L. G.,** Material requirements for circular magnetic domain devices, *IEEE Trans. Magn.*, MAG-5, 558, 1969.
18. **Wu, J. C., Hwang, J. P., and Humphreys, F. B.,** Operation of magnetic bubble logic devices, *IEEE Trans. Magn.*, MAG-20, 1093, 1096, 1984.
19. **Keyes, R. W.,** Cylindrical domains in anisotropic ferromagnets, *Comments Solid State Phys.*, 3, 135, 1971; reprinted in *Magnetic Bubble Technology,* Chang, H., Ed., IEEE Press, New York, 1975.
20. **Stroscio, J. A. and Eigler, D. M.,** Atomic and molecular manipulation with the scanning tunneling microscope, *Science,* 254, 1319, 1991.
21. **Eigler, D. M. and Schweizer, E. K.,** Positioning single atoms with a scanning tunneling microscope, *Nature (London),* 344, 524, 1990.
22. **Parthenopoulus, D. A. and Rentzepis, P. M.,** Three-dimensional optical storage memory, *Science,* 245, 843, 1989.
23. **Basche, T. and Moerner, W. E.,** Optical modification of a single impurity molecule in a solid, *Nature (London),* 355, 335, 1992.
24. **Bennett, C. H.,** The thermodynamics of computation, *Int. J. Theor. Phys.*, 21, 905, 1982.
25. **Conrad, M.,** Molecular computing as a link between biological and physical theory, *J. Theor. Biol.*, 98, 239, 1982.
26. **Wald, G.,** Molecular basis of visual excitation, *Science,* 162, 230, 1968.
27. **Govindjee and Govindjee, G. R.,** The absorption of light in photosynthesis, *Sci. Am.*, 231(6), 68, 1974.
28. **Clementi, E.,** Supercomputers and supercomputing in chemical research, *Intelligent Instrum. Comput.*, 7, 245, 1989; 8, 14, 1990.
29. **Aviram, A.,** Molecules for memory, logic, and amplification, *J. Am. Chem. Soc.*, 110, 5687, 1988.
30. **Meindl, J. D.,** Ultra-large scale integration, *IEEE Trans. Electron Devices,* ED-31, 1555, 1984.
31. **Heller, W. R., Donath, W. E., and Mikhail, W. F.,** Prediction of wiring space requirements for LSI, *Proc. 14th Design Automation Conf.*, New Orleans, 1977, 32.
32. **von Neumann, J.,** *The Computer and the Brain,* Yale University Press, New Haven, 1958.

Chapter 2

TOWARD POLYSILANE-BASED MOLECULAR ELECTRONIC DEVICES

Aleksander Herman

TABLE OF CONTENTS

0-8493-8061-8/93/$0.00 + $.50

I. COMPUTERS: PAST, PRESENT, AND FUTURE

Although the first electronic computers were developed during the early 1940s, the origins of computer history go back much earlier. In 1642 Blaise Pascal (1623–1662), after whom the Pascal programming language is named, developed an adding machine that operated on a system of gears and wheels. Thus it speeded up calculations while simultaneously making them more accurate.

The first machine to embody many of the same ideas as the modern computer was the analytical engine, invented by an Englishman, Charles Babbage (1792–1872). It accepted punched cards for input, had a memory, and could be programmed. Although Babbage worked on the analytical engine for more than 40 years, he was unable to complete it because of inadequate mechanical technology.

World War II brought the need to quickly develop better aircraft designs and weapons. To meet this need, scientists and engineers developed the first computers. In England at Cambridge University, there was the electronic delay storage automatic computer (EDSAC). In the United States there were the electronic numerical integrator and calculator (ENIAC) and the MARK I. Completed in 1944, MARK I was the largest electromechanical computer ever constructed. It actually embodied some principles of Babbage's analytical engine, which had been conceived more than 100 years previously.

Both the ENIAC and the EDSAC were electronic computers and used vacuum tubes instead of relays. Thus the swift motion of electrons replaced the somewhat slow movements of electromechanical switches and relays. This increased the speed of calculations 1000 times.

The University of Pennsylvania was the scene of the development of the ENIAC, which was completed in 1946. It could perform 5000 arithmetic operations per second. It contained 18,000 vacuum tubes, weighed 30 t, and filled several rooms. The rumor is that the lights in Philadelphia dimmed whenever the ENIAC was turned on.

Although the transistor was invented in 1948, it was not fully used in computers until the late 1950s and early 1960s. Transistors replaced vacuum

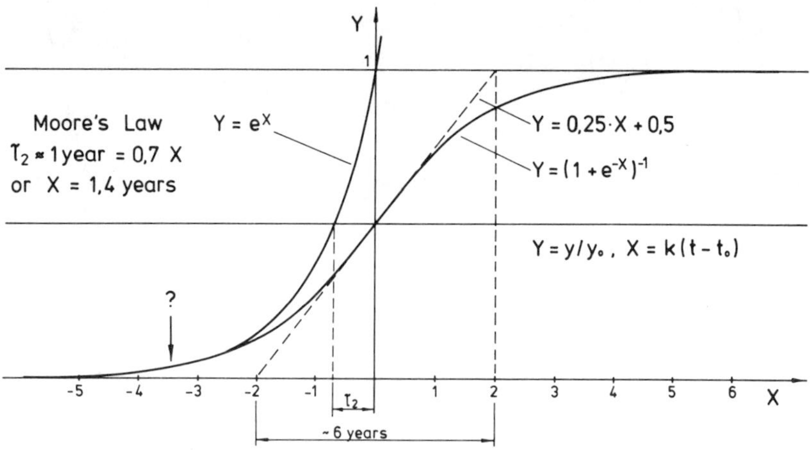

FIGURE 1. The simple mathematical model of the "logistic curve". From Moore's law we calculate the integration density doubling time $\tau_2 \approx 1$ year. The detailed discussion of the evolution of electronic information processing from small-scale integration to ultra-large-scale integration and beyond,[8] suggests that we are near the point indicated by the vertical arrow.

tubes, effecting an increase in the computer's processing speed, reducing the heat it produced, and greatly decreasing its size.

After transistors came integrated circuits. In an integrated circuit, transistors and related components were manufactured in a single circuit board. A computer might contain many of these boards. This new technology provided a decrease in computer size and an increase in speed.

The basic knowledge about history of computers can be found in many books and encyclopedias; our introductory information, for example, is taken from a popular book about FORTRAN programming by Smith,[1] in which interesting photographs from an IBM collection are shown.

Today integrated circuits are built on chips less than one-fourth of an inch square. The chips, made of a semiconductor material, can have access times less than a few nanoseconds. While the semiconductor very large scale integration (VLSI) chip is probably the most complex structure fabricated by man, it is interesting what might be achieved in the future.

It is well known that if you plot the results of any scientific or technological development against time, you obtain a kind of "S"-shaped curve called the "logistic curve".[2,3] The simplest mathematical model of the logistic curve is shown on Figure 1. For $e^{-X} >> 1$ ($X = k(t - t_0)$), the curve can be approximated by $Y = e^{k(t-t_0)}$. In the form of $\log Y = at + b$, it shows the famous Moore's law, which plots integration density chronologically.[4-8]

One can ask the question, at which point on the time scale are we now? If we are near $X = -2$, then exponential growth of integration density will change to a linear one, and after the next 6 years we will be near the limit

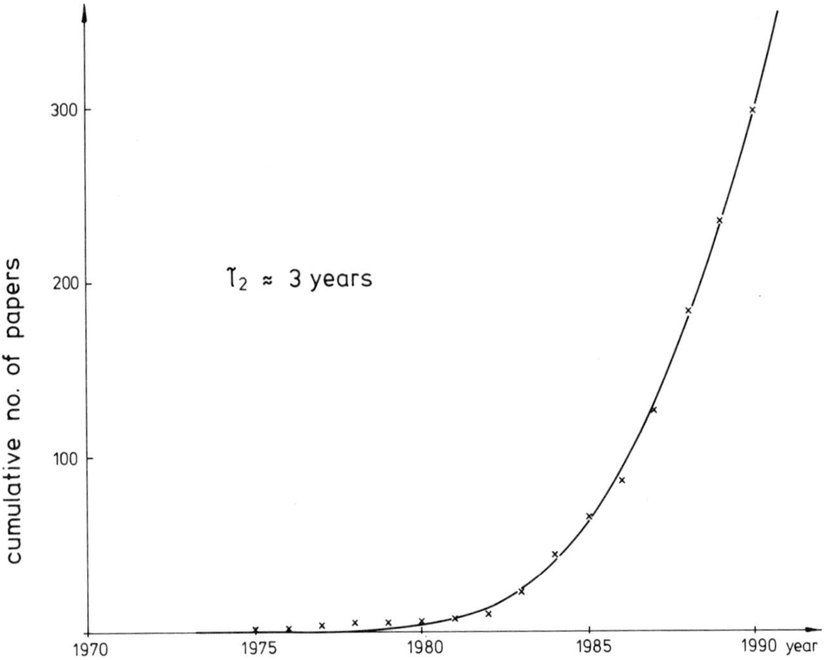

FIGURE 2. Cumulative number of papers related to molecular electronic devices and molecular electronics according to BIOSIS, CA, and INSPEC data files.

of ultralarge-scale integration (ULSI) technology at which level the conventional solid-state ideas will break down.[8] For example, at some hundred angstroms the linear dimension will become comparable to the mean free paths of electrons, phonons, and other excitations. When finally the linear dimension is only a few interatomic distances, the energy bands will split up into discrete levels. Therefore, there will be a discontinuity somewhere in the shrinking process.

It is impossible to answer this question perfectly, but probably we are now near the point when exponential growth of integration density will change to a linear one. Thus the suggestion that molecular materials offer the new course toward information processing, can open new possibilities and start a new logistic curve directly above the actual one based on silicon chips. This is the idea of "molecular electronics", a somewhat new discipline introduced by Aviram and Ratner.[9] Its advocates and adversaries have proclaimed extreme positions ranging from extolling the need to find the area at a level comparable to that enjoyed by the U.S. space program to claiming that it is totally unrealistic.[10-12] Independently on the results of this discussion, we just believe that it is an exciting intellectual adventure to see how far we can walk along this way. Figure 2 shows dynamics of discussion and preliminary research toward this objective, according to BIOSIS, Chemical Abstracts (CA), and INSPEC literature data files.

The field is very popular today ($\tau_2 \approx 3$ years) and our excitement might mislead the reader to expect a technological revolution during the next few years. If so, there will be a big disappointment when our reader learns that it will take us not years but decades to find out whether a data processing technology based on switching of individual molecules is a promising route to follow.[13]

A consequence of the wavelength of light and quantum fluctuations is that information cannot be reliably manipulated at a fast enough rate in an area significantly less than that expected from the evolution of semiconductor technology. Increased access speeds and lower heat generation, however, can be envisioned if electronic-based devices are superseded by optical-based devices. These could be molecule/polymer based. This would additionally enable the use of a wide optical bandwidth, i.e., the frequency domain, which would be equivalent to a substantial increase in the data density — an objective of molecular electronics. Thus, molecular electronics for serial applications will be best achieved through the discovery, exploration, and exploitation of bistable materials, which are switchable by light. Primary examples include spectral hole burning and nonlinear optical properties of materials.[10-12] New materials are key to molecular electronics, and the opportunities for synthetic chemists are vast.

This chapter of the book is not a report on experimental results obtained for polysilane-based molecular electronic devices. It is not a review of present application ideas of polysilanes as materials in molecular electronics either, since only few such ideas[14] have been reported in literature according to BIOSIS, CA, and INSPEC data files. The purpose of this contribution is to stimulate discussions, which in some future may lead to a selection of properties of polysilane materials for either serial/digital or preferably non-von Neumann (parallel/analog) applications. This does not imply that a polysilane approach to highly integrated electronic devices will be possible. The author just believes that the search for materials for future electronic applications in the rapidly growing field of polysilane chemistry (Figure 3) has led to the development of new technologies and commercialized products,[15] and that more examples are expected in the future.

Designing an electronic device that would operate at the molecular level involves several steps. It is first necessary to imagine a molecule or series of molecules that may possess the desired features, to synthesize particular examples, and to study their properties. At this step, it is crucial to have a simple theory that links the structure with properties of considered molecules. The second step is to incorporate it into supramolecular architectures, such as membranes or other organized structures, and to investigate whether the resulting entity possesses the required properties. This is the field of supramolecular chemistry.[16] The third step requires the connection of the basic unit

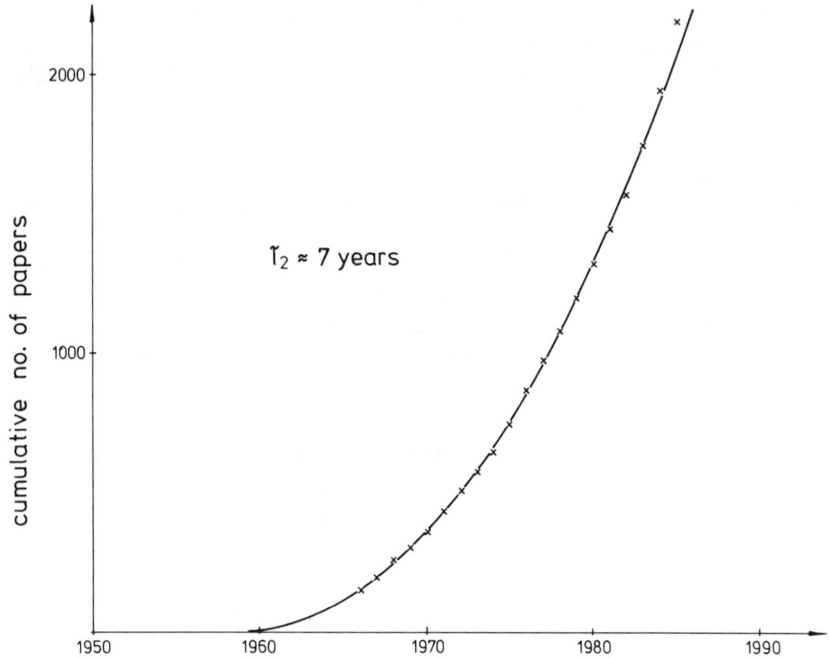

FIGURE 3. Cumulative number of papers about polysilane chemistry according to CA data file.

to other components, in order to address it via relay molecules or with an external physical signal. Of course, the goal of designing a molecular electronic circuit based on polysilane materials may have many spin-offs along the way, besides raising novel questions about the handling of polysilanes. It is probably premature to try to define clearly the routes to the goal: indeed, the definition of the goal itself may change under the influence of progress made along this path.

II. THE σ-CONJUGATED POLYSILANE POLYMERS

Polysilanes are compounds that contain silicon-silicon bonds in rings or linear chains. The oligosilanes and polysilane high polymers with up to more than 40,000 silicon atoms have been isolated within recent years.[17] As comprehensive summaries about polysilane chemistry and properties of polysilanes have appeared recently,[15,17] in this section only a simple theory of silicon backbone is introduced.

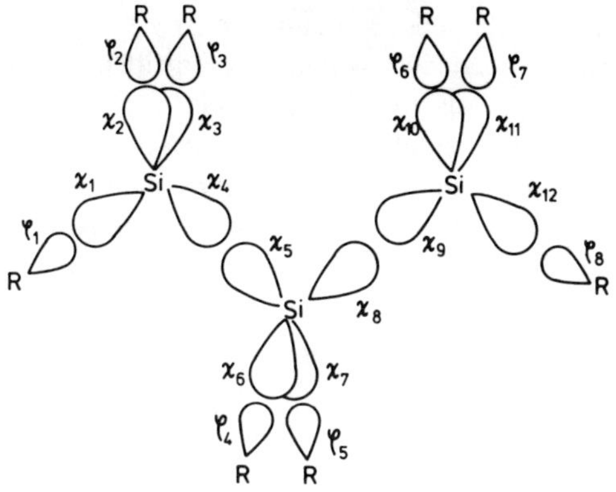

FIGURE 4. Trisilane bonding orbital description in terms of hybrid atomic orbitals (HAO) of silicon atoms χ_i and substituents φ_j. If χ_i orbitals are more diffuse than φ_j orbitals and lie considerably higher in energy, then the Sandorfy "C" approximation can be introduced and only χ_4, χ_5, χ_8, and χ_9 HAO considered explicitly (Set *A*). All other orbitals (Set *B*) form on approximately static potential for set *A*.

A. THE HYBRID ORBITALS INTERACTION MODEL OF THE SILICON BACKBONE AT VARIOUS LEVELS OF APPROXIMATION

For a long time chemists dealing with the polysilane family of molecules showed a rather marked preference for a model of bonding considering bond formation by interactions between hybrid atomic orbitals (HAOs) of silicon (χ_i, i = 1, . . . ,4) and atomic orbitals (AO) or HAOs of substituents (φ_j) (see Figure 4); each HAO (χ_i) interacted only with the corresponding ligand AO or HAO of adjacent silicon atom, the silicon HAOs being mutually orthogonal.[18]

All the semiempirical methods considered here assume that it is sufficient to take explicitly into account only some valence electrons, particularly those of the Si–Si backbone chain. This is an old idea, which originated from semiempirical calculations for alkanes,[19] for which the C–H bonds could be treated as localized two-electron systems. Such an assumption is not strange to chemists, for it is inherent to chemical intuition, which in turn follows from the observation that ionization energy of C–H bond remains almost unaffected in most alkanes. Thus analogously, the partition of valence electrons of polysilanes into localized Si–R bonds (R = alkyl, but not aryl) and delocalized Si–Si subsets to not contradict elementary chemical considerations and polysilane photo electron spectroscopy (PES) data interpretation.

From the quantum chemistry point of view, separation into electronic groups allows the total wave function Ψ to be represented as an antisymmetrized product of two or more subset wave functions:[20]

$$\Psi = A[\Psi_A \Psi_B \Psi_C \ldots] \tag{1}$$

The separability of wave functions of subset A, B, C . . . is obtained when the so-called strong orthogonality condition is fulfilled.[20,21] If, for example, subset A comprises π-type electrons, and subset B comprises σ-type electrons the strong orthogonality condition is automatically satisfied on symmetry grounds, and is known as the σ–π separability condition. For the σ_{SiSi}–σ_{SiR} electron separation, orthogonality can be forced artificially by using orthogonal orbitals for both sets.

For all valence electrons of a given system the variational equations analogous to the Hartree-Fock (HF) equations can be derived assuming the separability of subsets A and B. One subset, for example, A, can represent the Si–Si bonding electrons; and the form of the one-electron Hamiltonian will then change to the effective one. For subset A this Hamiltonian will be given by:

$$\hat{H}^{eff,A} = \sum_{k=1}^{n_A} \hat{h}^{eff,A}(k) + \frac{1}{2} \sum_{\substack{k,l \in A \\ k \neq l}} \hat{g}(k,l) \tag{2}$$

where n_A is the number of electrons in subset A, and:

$$\hat{h}^{eff,A}(k) = \hat{h}(k) + (\hat{J}^B(k) - \hat{K}^B(k)) \tag{3}$$

The one-electron term $\hat{h}^{eff,A}$ contains an additional potential energy operator that accounts for the presence of the subset B. Therefore, semiempirical calculations in the Sandorfy "C" approximation use the effective one-electron Hamiltonian (Equation 3). The corresponding scheme will involve matrix elements of the form $\langle \chi_i | \hat{h}^{eff,A} | \chi_j \rangle$ where χ_i and χ_j are orbitals of the subset A. These are frequently treated as empirical parameters. The second fundamental assumption is the zero differential overlap approximation.

The description of bonding presented above can be used both as: (1) a qualitative method of analysis (level pattern indices [LPI], electron count indices [ECI], structural covariance [SC], and deformational covariance [DC])[22-29] and (2) for Hückel-like semiquantitative simple linear combination of valency points (LCVP) molecular orbital (MO) calculations[30-34] also. The latter is used in the sense of "best" semiempirical HF self-consistent field configuration interaction [SCF CI(1)] MOs[35-37] through Christoffersen's *ab initio* method,[40] as the essential starting point of the generalized valence bond (GVB) method,[41-47] which includes the electron correlation effects lacking in MO theory.

1. Sinanoğlu's Valency-Points Interaction Approach

A pictorial "blackboard mnemonic" type method of handling the crucial electronic property of molecules introduced recently by Sinanoğlu[28,29] allows the molecular orbital level patterns and also the number of nonbonding,

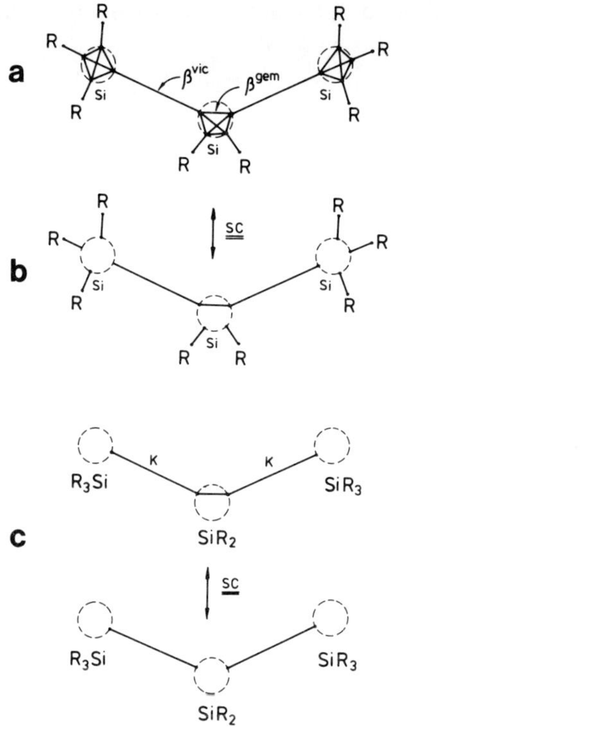

FIGURE 5. Valency points interaction formulas of trisilane. (a) Full description; (b) after the Sandorfy "C" approximation; (c) as above but without localized Si–R interaction lines, and with assumption that $\beta^{gem} \equiv 1$ and $\kappa = \beta^{vic}/\beta^{gem}$. In fact, $\beta^{gem}/\beta^{vic} = 0.87 \pm 0.01$ for polysilanes.[31,32] The ($\overset{SC}{=}$) lines between VIFs indicate the latter are "structurally covariant", having therefore the same MO level patterns, LPI = {number of bonding (n_+), nonbonding (n_0), antibonding (n_-) MOs}. LPI of trisilane = {$n_+ = n_- = 2$}.

bonding, and antibonding orbitals to be calculated from the actual or tentative structural formulas of molecules or intermediates regardless of symmetry.

According to Sinanoğlu, each HAO considered represents a valency point (VP) in valency-points interaction formula (VIF) of a molecule, which contained all VP-VP′ interaction lines (VL) between interacting VPs (Figure 5). Nearest-neighbor VPs, for example, will in general interact.

Each molecular structure corresponds to a one-electron effective Hamiltonian, $\hat{H}^{eff,A}$, which is parametrically dependent on the three-dimensional atomic positions in {**R**}. In fact, the VIF depicts directly this operator $\hat{H}^{eff,A}$ with χ_i being a VP, and $\beta_{ij} = \langle\chi_i|\hat{h}^{eff,A}|\chi_j\rangle$ being a line between the VPs i and j. VPs also have self-energies $\beta_{ii} \equiv \alpha_i$. Where α's of different VPs differ greatly, there also will be "loopline" j of strength $\xi_j = (\alpha_j - \alpha_0)/\beta^0$ in the VIF, where β^0 is the "standard" β_{ij} line strength and α_0 is the standard VP self-energy.

With pure AO-type VPs there are no intra-atomic lines since:

$$\beta_{ij}^{gem} = \langle \chi_{ia} | \hat{h}^{eff,A} | \chi_{ja} \rangle = 0$$

in such a case. With HAO's

$$\beta_{ij}^{gem} = \langle \chi_{ia} | \hat{h}^{eff,A} | \chi_{ja} \rangle \neq 0$$

in general. The relationship between β_{ij}^{gem} value and hybridization of the atom has been known for many years,[48,49] but its chemical significance has been appreciated only recently by Dewar.[50-52] For example, the β^{gem} integral between two sp^3 hybrids is equal to $(I_s - I_p)/4$, for two sp^2 hybrids is equal to $(I_s - I_p)/3$, and for sp hybrids is equal to $(I_s - I_p)/2$, where I_s and I_p are the valence state ionization energies of *s* and *p* AOs, respectively.

The LPI to be deduced from the VIF is independent of the actual magnitude of the line strengths (ξ_i and $\kappa_{ij} = \beta^{ij}/\beta_0$) often within a wide range of those. This range can be found from the VIF by the two VIF rules.[28,29]

Rule 1: Any valency point VP of a VIF may be multiplied by an arbitrary constant κ, positive or negative while it is not zero. This changes, distorts, etc. the VIF but does not change its LPI. Multiplying a VP by κ means multiplying the strengths of all the lines emanating from that VP by κ.

Rule 2: Any VP of a VIF may be lifted up, taken over, and placed onto any other VP′ carrying along the lines that terminated on VP onto the new VP′ that becomes a terminus of those lines. If in the process two lines superimpose, their strengths are added algebraically giving the strength of new line. The original lines that terminated at the initial VP and that VP are also retained in the resulting VIF′. Then VIF and VIF′ will have the same LPI.

There are "critical" values of some ξ and κ, however, where suddenly the LPI changes. These values play a crucial role in determining when a molecule rearranges, changes hybridization type on some of its atoms, etc.

With two VIF rules one can generate all the molecules (or reacting assemblies of molecules) that are iso-LPI, therefore of similar energetic quality (if ECI also does not change, i.e., the rearrangement or reaction is electronically adiabatic). Using the rules on a VIF the LPI itself also can be obtained.

2. Simple Hückel-Like LCVP MO Calculations

Although the oldest one, the simple Hückel-like method is a refinement of the Sinanoğlu approach. It uses two orbital interaction parameters: β^{vic} for interacting orbitals of vicinal VPs and β^{gem} for the VPs on the same atom. Each VP of silicon is assumed to have the same self-energy α_i and substituent and end effects are absorbed by the parametrization unless additional parameters are introduced.[31,32]

The solution of the secular determinant gives eigenvalues and eigenvectors coefficients for the basic orbitals. It is then possible to calculate orbital electronic charges, polarizabilities, and electronic energies quite analogous to those calculated in applications of the Hückel π MO theory.

The optimal values of the α and β parameters have been determined by fitting the experimental IPs of a catenate series of molecules to solutions of the secular equation.[31,32] Using parameters derived in such a way, the ability of this simple model to correlate and predict the properties of the other members of catenated molecules is surprisingly good. For example, the standard error of calculations of the first and higher IPs of a member of polysilanes is 0.111 eV: that is almost equal to the experimental uncertainty, which is up to 0.10 eV.[30]

A similar fitting of α and β parameters to experimental values of transition energies of linear polysilanes unfortunately yields a different set of parameters that reproduce experimental data with comparable accuracy.[18] Separate parameter sets exist for separate problems, i.e., Hückel-like methods do not yield accurate results for different classes of observables with one set of parameters. Thus the method, the parameters, and the results are often treated with suspicion. In the next part we would like to present an internally consistent model that leads to results that may be used with confidence.

3. Advanced Semiempirical LCVP SCF CI(1) MO Method

Similar to the above considerations it is assumed that σ_{SiSi} electrons of the polysilane conjugated system constitute a separate electronic group. The atomic nuclei inner-shell electrons and electrons of σ_{SiC} and σ_{CH} bonds only define a static potential field for these σ_{SiSi} electrons. Thus, the core Hamiltonian has the form of Equation 2. The many electron wave function for even number n_A, σ_{SiSi} electrons in a closed-shell state is represented by a single Slater determinant built of σ_{SiSi} molecular orbitals, which are in turn determined from the HF equations essentially similar to the original Pariser-Parr-Pople (PPP) theory.[35,36]

Matrix elements of the HF operator are calculated as follows:

1. The atomic basis comprises VPs that contribute σ_{SiSi} electrons. For simplicity we assume that each VP contains the same part of sp mixture, i.e., they are equivalent and orthonormal:

$$\langle \chi_i | \chi_j \rangle = \delta_{ij} \tag{4}$$

   The Gaussian-like orbitals χ are placed on barycenters of hybridized sp^3 Slater-type orbitals of silicon atom in tetrahedral configuration.
2. The zero differential overlap (ZDO) approximation is then introduced for all VP matrix elements except the core.

3. The nearest-neighbor approximation is introduced for nondiagonal elements of the core Hamiltonian, i.e.:

$$H_{ij} = \begin{cases} B_{ij} & \text{for i,j being the nearest neighbors} \\ 0 & \text{otherwise} \end{cases} \quad (5)$$

These assumptions result in the following HF matrix elements:

$$F_{ij} = \begin{cases} \omega_i + \frac{1}{2} P_{ii}\gamma_{ii} + \sum_{a \neq i} (P_{ii} - 1)\gamma_{ai} & \text{for i = j} \\ \beta_{ij} - \frac{1}{2} P_{ij}\gamma_{ij} & \text{for i,j being the nearest neighbors} \\ -\frac{1}{2} P_{ij}\gamma_{ij} & \text{otherwise} \end{cases} \quad (6)$$

where P_{ii}, P_{ij} are the elements of the σ electron charge density-bond order matrix P.

The penetration integrals are averaged and included in ω_i's. Core integrals ω_i, β_{ij}^{gem}, and β_{ij}^{vic} are treated as empirical parameters. The one-center two-electron integrals are approximated by the difference between the orbital ionization potential and the orbital electron affinity for sp^3 orbitals. The two-center integrals β_{ij}^{vic} are calculated using an empirical formula for their dependence on intercenter distance. For two-electron two-center repulsion integrals the modified formulas of Mataga and Nishimoto are used, with distance r_{ij} calculated between the centers of χ_i and χ_j orbitals. The integrals γ_{ij}^{gem} are treated as an empirical parameter.

The advanced semiempirical LCVP SCF CI(1) MO method was introduced in 1986[35] despite the existence of several more sophisticated approaches that are mostly more computer time-consuming. The parameters of this method have been mastered by fitting to the experimental spectroscopic PES and UV data of dimethylsilylene (Me_2Si:) and hexamethyldisilane ($Me_3SiSiMe_3$), so that more refined schemes cannot compete in accuracy with ionization and transition energies calculated by this method.

Sandorfy and Daudel's dissection of C–C bonds from the C–H bonds in alkanes[19] has made a profound contribution to our understanding of the physical and chemical properties of polysilanes. It has been the prime objective of this account to discuss the results of recent adaptations of a similar approximation to polysilanes, carried out to study their properties. There seems to be no doubt that the idea mentioned above could be a useful starting point for exploration of polysilane chemistry and physics. The model may be improved by allowing the π basis MO to mix with σ_{SiSi} MOs.[38] A good agreement between experimental data and the model is obtained through consideration of the mixing of the π basis MO only with the σ_{SiSi} MO in phenyl-substituted methylpolysilanes. Software for such calculations is now available.[39]

4. Christoffersen's *ab initio* Method

The first step in this method[40] consists of examination of small molecules or molecular fragments that appear in larger systems. The aim is to obtain an adequate description of the electron distribution in these fragments using as small a basis set as possible, but not restricting the basis orbitals to lie on nuclei. In this procedure, normalized floating spherical Gaussian orbitals (FSGO) are used as the basis orbitals for description of VPs or isolated VLs. The FSGOs are defined as:

$$G_i(r) = (2\pi/\rho_i^2)^{3/4} \exp\{-(\mathbf{r} - \mathbf{R}_i)^2/\rho_i^2\} \tag{7}$$

where ρ_i determines the size of the FSGO (and is often referred to as the orbital radius), and $\mathbf{R}_i$ is the location of the FSGO, relative to some arbitrary origin. Of course similar to the semiempirical approach, many criteria are possible for judgment of the adequacy of the fragment description. In the Christoffersen approach, some consideration is given to the energetics of various fragment descriptions, but it should be stressed that the ultimate decision on the acceptability of a fragment description is obtained by an examination of small- and medium-sized prototype molecules containing the fragment. It is of interest to note that, in the absence of symmetry, there are four nonlinear parameters to be determined for each FSGO. This restricts the practical utility of the procedure to descriptions involving only a few FSGO. For molecular fragments that consist of a closed-shell system of 2N electrons, the trial wave function is taken to be a single, normalized Slater determinant of doubly occupied, nonorthogonal orbitals, i.e.:

$$\Psi(1,2,\ldots,2N) = \{[(2N)!]\det\Delta\}^{-1/2}\det\{\chi_1(1)\alpha(1)\chi_1(2)\beta(2) \cdots \chi_N(2N)\beta(2N)\} \tag{8}$$

where:

$$(\Delta)_{ij} = \int \chi_i^*\chi_j \, dv \tag{9}$$

In some studies, χ_i will be taken to be a single FSGO. In other studies, a "split FSGO" description will be used, in which a linear combination of two FSGO is used for a given χ_i. Here, both the linear and nonlinear parameters are included in the parameter variations. Neither of these choices represents restrictions on the method, but rather, is convenient for initial investigations. In fact, they illustrate the case in which the procedures can be modified to improve the accuracy of the description, if needed.

The other type of molecular fragment that arises contains a single unpaired electron, and the trial wave function for this open-shell system is assumed to be in the form:

$$\Psi(1,2,\ldots,2N + 1) = \{[(2N + 1)!]\det\Delta\}^{-1/2}$$

$$\det\left\{\chi_1(1)\alpha(1) \cdots \chi_N(2N)\beta(2N)\chi_{N+1}(2N + 1) \begin{matrix}\alpha(2N + 1)\\ \beta(2N + 1)\end{matrix}\right\} \quad (10)$$

where the (N + 1)st orbital may have either α or β spin.

The general expression for the nonrelativistic electronic energy that results from choices of trial wave functions can be written as:

$$E_{el} = 2\sum_{i=1}^{N+P} \langle i|h|i\rangle - \delta_{P,1}\langle N + 1|h|N + 1\rangle + \sum_{i=1}^{N+P}\sum_{j=1}^{N} \{2\langle ii|jj\rangle - \langle ij|ji\rangle\} \quad (11)$$

where

$$P = \begin{cases} 0 & \text{for the 2N electron case} \\ 1 & \text{for the (2N + 1) electron case} \end{cases} \quad (12)$$

$$h = -\frac{1}{2}\nabla^2 - \sum_{\nu=1}^{\text{all nuclei}} Z_\nu/r_\nu \quad (13)$$

and

$$\langle ij|kl\rangle = \iint \sigma_i^*(1)\sigma_j(1)r_{12}^{-1}\sigma_k^*(2)\sigma_l(2)dV_1\, dV_2 \quad (14)$$

where σ_i is a member of an orthonormal basis set. The current method uses nonorthogonal basis set χ, which can be transformed into an orthonormal basis set Ω by means of:

$$\Omega = \chi\Delta^{-1/2} \quad (15)$$

A list of the molecular fragments that have been found to be useful in the description of polysilanes is given in Table 1.

Several comments are appropriate regarding the entries of Table 1. First, it is interesting that only a few silanes are needed to form an enormous number of polysilanes of interest to us. Second, the number of independent nonlinear parameters that must be varied is reasonable and tractable; however, it should

TABLE 1
Polysilane Fragments of Interest in Eight FSGO Descriptions

Polysilane fragment (symmetry)	No. of independent nonlinear parameters
SiH_4 (T_d)	4
$RSiH_3$ (C_{3v})	10
R_2SiH_2 (C_{2v})	12
$RR'SiH_2$ (C_s)	18
R_3SiH (C_{3v})	10
RR'_2SiH (C_s)	18
$RR'R''SiH$ (C_1)	26

Note: The description corresponds to the use of four FSGO for each inner shell on silicon atom and four FSGO for bonding electron pairs; local tetrahedral symmetry is assumed. An alternative or additional description can be obtained by ''splitting'' FSGO. The description of substituent orbitals is omitted for simplicity.

be noted that simple extension of this notion to oligosilanes is not possible, since the simultaneous determination of both linear and nonlinear parameters for a large molecule of low symmetry is not possible.

Third, it is not necessary that the molecular fragments correspond to actual stable molecular species, since they are merely useful constructs that approximately describe a variety of anticipated molecular environments. Thus the bond distances and angles chosen for fragment studies do not necessarily correspond to the values observed for the actual molecule. The potential field in which the basis orbitals are determined is a molecular field (as opposed to the use of atomic calculations for this purpose), with hydrogen atoms serving as an approximation to the environment expected when new bonds are formed. In the determination of optimum parameters for fragments containing silicon atoms a computational difficulty arises with FSGO core representation (see Figure 6).

If free variation of all parameters is allowed, the FSGO describing the tetrahedral core orbitals will coalesce onto the silicon nucleus. Thus, these FSGO positions need to be chosen using other considerations. The analysis and numerical studies of Spangler and Christoffersen[53] show that satisfactory accuracy in integral evaluation can be maintained if the ratio of distance from the origin to orbital radius is taken to be in the range 0.002 to 0.005 for computers having 72-b words, or in the range 0.005 to 0.01 for computers having 64-b words. Since the interpretation of FSGO results is always given in terms of valence electron behavior and since the FSGO core representation does not enter directly into the valence description, a pseudopotential procedure for use in FSGO calculations has been developed.[54]

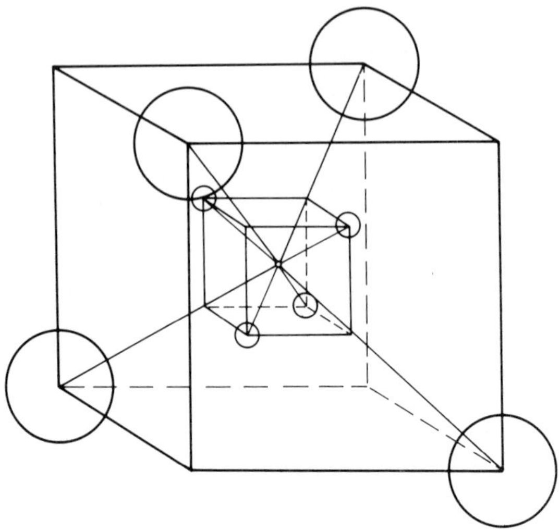

FIGURE 6. The FSGO basis set of the silicon atom. In the argon configuration of SiH_4, $RSiH_3$, etc. we could think of an outer octet and inner octet of electrons. These differ in energy and in spatial extent by an order of magnitude, but we can think of the outer octet distribution as forming a perturbing negative potential for the inner octet. If the outer octet is concentrated into one set of tetrahedral directions, it should push the inner octet toward the opposite set of directions as shown in the figure. The inner shell repulsions between two such directed electron clouds may have a bearing on the difficulty of double-bond formation between third row and heavier atoms.

After the description of the appropriate molecular fragments has been completed, the fragments and their associated FSGO are combined appropriately to form polysilane molecules. This procedure can be considered conveniently as consisting of several steps:

1. Bring the appropriate molecular fragments and their associated FSGO together to the desired distances and angles.
2. Remove the extra hydrogen atoms (including their electron) from the Hamiltonian. However, retain the FSGO that were used to describe Si–H bonds, that will now describe a newly formed Si–Si bond (a pair of FSGO for each new bond or in Sinanoğlu's terminology a pair of [VPs] for each new [VL]).
3. Obtain an optimum molecular orbital description for the molecule for this basis set by the solution of the ordinary SCF equations.

In particular, the molecular orbitals can be written as:

$$\psi_i = \sum_{A=1}^{P} \sum_{k=1}^{N_A} c_{ki}^{A} \chi_k^{A} \tag{16}$$

$$\left\{ G_1^A\ G_2^A \cdots G_8^A\ ;\ G_1^B\ G_2^B \cdots G_8^B \right\}$$

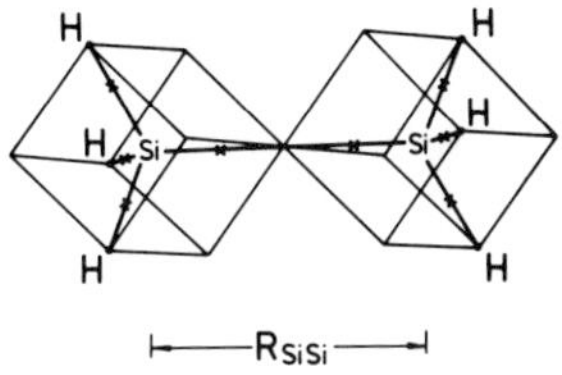

$$\psi_i = \sum_{P=A,B} \sum_{k=1}^{8} C_{ki}^P G_k^P$$

FIGURE 7. Formation of disilane from two SiH_4 fragments; × denotes the approximate position of the outer (valence) FSGO. The inner FSGOs are omitted for simplicity.

where the χ_k^A are the previously determined fragment orbitals ($\chi_k^A \equiv G_k^A$ for most studies discussed here); and the summations are taken over all fragments P and all orbitals within a fragment N_A, opposite to semiempirical treatments where only Si–Si bonding hybrids are taken into consideration when the Sandorfy "C" approximation is introduced. Thus, the changes in electronic and molecular structure that are caused by formation of Si–Si bonds are reflected in the MO coefficients (c_{ki}).

The above procedure is illustrated for the formation of disilane from two SiH_4 fragments in Figure 7.

As shown in the figure the basis set is no longer a minimum basis set, and the SCF procedure will improve the energy, causing one virtual orbital and 15 filled orbitals here. One also should notice the role of the hydrogen atoms more explicitly through this figure. It should be clear from this example that one or more of the hydrogen atoms in a molecular fragment may be removed, with the formation of an additional bond with each removal.

For the closed-shell molecules of interest in these studies, the wave function determined by the SCF procedure has the form:

$$\Psi(1,2,\ldots,2N) = [(2N)!]^{-1/2}\det\{\psi_1(1)\alpha(1)\psi_1(2)\beta(2) \cdots \psi_N(2N)\beta(2N)\} \quad (17)$$

and the SCF equations to be solved are given by:

$$\mathbf{Fc}_i = \Delta \mathbf{c}_i \epsilon$$

$$(\mathbf{F})_{rs}^{AB} = \langle \chi_r^A | h | \chi_s^B \rangle + \quad (18)$$

$$\sum_{C=1}^{Q} \sum_{D=1}^{Q} \sum_{t=1}^{N_C} \sum_{u=1}^{N_D} P_{tu}^{CD} \{ \langle \chi_r^A \chi_s^B | \chi_t^C \chi_u^D \rangle - 1/2 \langle \chi_r^A \chi_u^D | \chi_t^C \chi_s^B \rangle \} \quad (19)$$

and ϵ is a diagonal matrix of orbital eigenvalues. Also, **P** is the charge and the bond order matrix, defined as:

$$\mathbf{P}_{tu}^{CD} = 2 \sum_{j=1}^{occ} c_{tj}^{C} c_{uj}^{D} \tag{20}$$

One of the striking features of the *ab initio* calculations on organopolysilanes that have been carried out to date is that investigations at the "large" end of the large molecule scale (larger than permethyltetrasilane[55]) are practically nonexistent. Thus, the practical applicability of Christoffersen's method to deal with polysilanes must be considered as an effort that has been made to develop an *ab initio* procedure to which comparison of semiempirical schemes described above may be possible.

B. ELECTRONIC STRUCTURE AND PROPERTIES OF POLYSILANES

Although the first organosilane polymers (those having only silicon atoms in the backbone) were reported in 1924,[57] they evoked only scientific interest because of their insolubility and intractability. With the arrival of the synthetic materials era in science, the situation changed radically. Since 1975 we have witnessed the dawn of what one might call the potential multifaceted application epoch in polysilane chemistry. These polymers have attracted considerable interest in applications such as precursors to β–SiC fibers, impregnating agents for ceramics, photoresistors, photoinitiators for vinyl polymerization, semiconductors upon doping, etc. The "publication explosion" in this field in the late 1970s (Figure 3) has induced Miller and Michl[15] to write the review in which relevant literature is surveyed through 1988.

On the other hand, while large-scale computer calculations may provide data on selected polysilane species, day-to-day reasoning requires easy to use qualitative- or semiquantitative-type easy to use methods with large numbers and kinds of polysilanes. Thus, the abilities of the methods introduced above to describe characteristic properties of polysilanes are investigated, and some experimental effects observed in polysilane chemistry and physics are interpreted according to hybrid orbital interaction models at various levels of approximation.

1. Some Aspects of Polysilane Synthesis

The condensation reaction of dichlorosilanes with alkali metals still gives the main practical route to cyclic polysilanes.[15] Two features of heterogeneous polycondensation become obvious upon examination of the experimental data:

1. Polysilane yields in a solvent such as toluene decrease markedly with increasing stearic bulk of the substituents.
2. Many polycondensation reactions are strongly solvent dependent; experimentation often results in greatly improved yields of high polymers.

3. Molecular weight distributions in the crude polymer mixture are often strongly polymodal.

A trimodal distribution is often produced with contributions clustered in three distinct molecular weight regions:[15] mol wt < 1,500; mol wt = 4,000 to 30,000; and mol wt = 100,000 to several million. The material in the lowest molecular weight region is mainly oligomeric and interesting for mechanistic studies of the condensation reaction.

Some years ago we proposed[32] a mechanism of this reaction according to which bifunctional linear species of the type $M(R_2Si)_nX$ are formed as intermediates, where M = alkali metal and X = halogen. The chain formation, growth, and cyclization take place in three competitive-consecutive reactions:

$$X(R_2Si)_nX + 2M \xrightarrow{k'_n} X(R_2Si)_nM + MX \tag{21}$$

$$X(R_2Si)_nM + X(R_2Si)_mX \xrightarrow{k_{nm}} X(R_2Si)_{n+m}X + MX \tag{22}$$

$$X(R_2Si)_nM \xrightarrow{k_n} (R_2Si)_n + MX \tag{23}$$

The rate of the first reaction (Equation 21) must be very much alkali metal surface dependent. The second and third step (Equations 22 and 23) are competitive reactions. The rate of bimolecular step (Equation 22) must be very strongly dependent on concentration, and the rate of unimolecular process (Equation 23) should be only dependent on chain length (n).

Simple Hückel-like calculations based on VPs and the Sandorfy "C" approximation with PES-scaled values of integrals predict that no electronic activation barriers are present for cyclization of odd-membered homologues according to the reaction in Equation 23. This result agrees with the observation that five- and seven-membered rings are produced by a kinetically controlled process. Calculations also show that this through-bond interaction control probably will be kinetically undetectable if more than seven σ_{SiSi} bonds are involved, since the computed activation barrier rapidly tends to zero as the number of silicon atoms grows to infinity.

Our experiments[32] with 1,2-dichlorotetramethyldisilane and 1,3-dichlorohexamethyltrisilane used as substrates for polysilane synthesis suggest that the so-called silylene mechanism of the polysilane formation cannot be ruled out, but probably operates much slower in comparison to that proposed by us (for n = 1, Equation 23 gives diorganosilylene, which can produce polysilylenes [linear polysilanes] by polymerization or insertion into Si–Si bonds). This kind of reaction probably can be more important for application toward polysilane-based molecular electronic devices since many diorganosilylenes are easy to obtain from its precursors by the way of thermo- or photolysis.[56]

2. Ground Singlet State Electronic Structure of Polysilanes and Its Relevance to Experimental Properties

The systems concerning molecular electronics most often discussed are conjugated polymers with alternating single and double bonds. Polysilanes represented the first example of molecules without π, d, or lone-pair (n) electrons, but showing very similar properties as organic conjugated polymers (for example, semiconducting properties due to doping). Experimental studies of electronic structures of conjugated polymers have been carried out mainly by means of X-ray photoelectron spectroscopy (XPS). UV photoelectron spectroscopy (UPS) is a more suitable tool for examining the upper most valence bands, which are most important when discussing the electric conduction; this is mainly because of its good resolution (~0.1 eV) and high sensitivity. Unfortunately, its application to polymers is rather limited. Thus the information on electronic structure of polymers must be drawn from UPS data concerning oligomers by extrapolation according to the one-electron molecular orbital theory. This part of the contribution reports those aspects of UPS parametrized molecular orbital theory that have the main impact on factors governing σ systems conjugation. The main concern is the applicability of conventional one-electron ideas to questions concerning transport properties of linear σ systems.

The dynamic properties of an electron moving along a polymer chain are described in the band theory by the effective mass idea. This idea presumes that electrons and holes localized in a momentum space near the band *extrema* can be treated as classical particles whose kinetic energy depends quadratically on their quasimomentum, $\hbar k$. Elementary Newtonian mechanics applied to such particles leads to the well-known expression for the effective mass:

$$m^* = \hbar^2\left(\frac{d^2E}{dk^2}\right)^{-1} \tag{24}$$

where E is the electron energy. Classical particles of this mass, moving in an electric field and subjected to damping by scattering processes, will have the conductivity given by:

$$\sigma = Ne^2\tau/m^* \tag{25}$$

where N is the total number of free carriers and τ is the scattering form rate, dependent on chain vibrations and defects.

TABLE 2
The Results of the Regressional Analysis of UPS Spectra of the Studied σ Systems According to Equation 26[30,60,61]

System	U [eV]	$\hbar/2m^*$[eV Å^2]	m*/m	I^{th} [eV]
$(CH_2)_n$	−13.021 ± 0.025	0.882 ± 0.014	4.320	9.35 (8.5)[a]
$(SiH_2)_n$	−10.997 ± 0.037	1.390 ± 0.058	2.741	8.49
$(SiMe_2)_n$	−9.426 ± 0.042	1.484 ± 0.071	2.566	6.75 (5.9)[a]
$(CH)_n$[b]	−12.209 ± 0.072	5.135 ± 0.111	0.740 (0.5)[c]	5.72 (5.24)[a]

[a] In parentheses literature data.
[b] The classical π electron system for comparison.
[c] For comparison theoretical value in parentheses.

The effective electron mass evaluation, based on UPS spectra was carried out under the following assumptions:[30]

1. In a molecule $R(R_2M)_nR$, only the M–M σ bond molecular orbitals are considered.
2. The electrons are treated as electrons in a one-dimensional potential well of length L = nD and depth U, where D is the average M–M bond distance.
3. The introduced effective electron mass m* is responsible for the M–R tails and electron-nucleus interactions.
4. m*/m and U are treated as scaling factors to reproduce the absolute separation of UPS energy levels according to Koopmans' theorem and equation:

$$VIP_j = -\epsilon_j = U - (\hbar k_j^2/2m^*) \qquad (26)$$

where $k_j = 2\pi/\lambda_j$ and $\lambda_j = 2L/n_j$; λ_j is the de Broglie wavelength of the jth free-electron orbital. (The parameters of Equation 26 can be determined using a standard least squares technique.)

Table 2 depicts the calculated values of the effective potential U, the effective electron mass m*, and the ionization energy threshold I^{th} for alkanes (M=C, R=H, and D = 1.54 Å); silanes (M=Si, R=H, and D = 2.34 Å); permethylsilanes (M=Si, R=Me, and D = 2.34 Å).[30] For comparison, the corresponding parameters for π orbitals of aromatic hydrocarbons are also included.

The differences between U values for series with M=C and M=Si may be interpreted by comparison with the valence state ionization energy of carbon and silicon atoms. The same interpretation applies to the difference between U values for σ systems of aliphatic hydrocarbons and π systems of aromatic hydrocarbons. For the series with the same M (M=Si) the U value depends on ionization energy of R. Despite absolute values, the striking difference between the m* values for σ and π systems of hydrocarbons should be emphasized. The m* values strongly depend on M, but only slightly on R. One can notice that the larger m* is, the smaller σ is (and thus the mobility of carriers). Finally, a rough estimate of the σ_{SiSi} curvature (another measure of the σ_{SiSi} electron effective mass near the highest occupied molecular orbital [HOMO]) also can be made (Equation 24); the numerical values of the variables that define the denominator of the effective mass, m*, are in the ratio polyethylene/polysilane = 1.58; the effective mass of polyethylene is thus 1.6 times more important than that of polysilane, a result agreeing with cell size effect. The ionization energy thresholds I^{th} were also calculated and compared with literature data, when possible. It can be concluded that polysilanes are the unique conjugated polymers with a very low well depth, and the carrier effective mass is an average between those of polyethylene and polyacetylene. Experimentally confirmed high conductivity of polysilanes after doping[15] should then result from the σ conjugation.

The delocalized electronic structure of the polysilane backbone perhaps can be most simply understood by reference to a pictorial "blackboard mnemonic"-type method of Sinanoğlu.[28,29]

According to this method, for any straight or branched polysilane chain, the VIF is reduced ($\overset{SC}{=}$) with Sinanoğlu's rules to a set of isolated Si–Si line segments similar to those demonstrated for alkanes.[28] The intrasilicon and intersilicon line strengths are of comparable magnitudes. However, even if their differences are taken into account, the reduction procedure by the two rules mentioned above reveals that the LPI is unaffected. Each simple line segment contributes $n_+ = 1$ and $n_- = 1$, so that for any such polysilane $n_+ = n_- =$ (number of Si–Si bonds). As there is one electron per VP in polysilanes, ECI = $\{N_+ = 2n_+, N_0 = N_- = 0\}$, where N_+, N_0, and N_- are the numbers of bonding, nonbonding, and antibonding electrons.

To compare the effect of σ bond conjugation on the electronic properties of polysilanes with the π bond conjugation in classical organic polymers, the valency points interaction formulas (VIFs) analogy is introduced at this point.

The relationship between the Hückel approximation, commonly used for π electron systems, and the Sandorfy "C" model used for σ systems is shown in Figure 8.

The figure depicts the analogy between two systems: the alternated polyene of $\beta_1 \geq \beta_2$ and a polysilane σ system with $\beta^{vic} \geq \beta^{gem}$. The analogy between both systems is perfect. The only difference is that polymeric orbitals consist of two p_z atomic orbitals of polyene carbon atoms for a π system, or approximately $3sp^3$ hybridized orbitals of backbone atoms for a σ system.

The energy calculation as a function of the reciprocal vector **k** for two orbitals per unit cell and a lattice vector of length gives a well-known solution with two energy bands:[58,59]

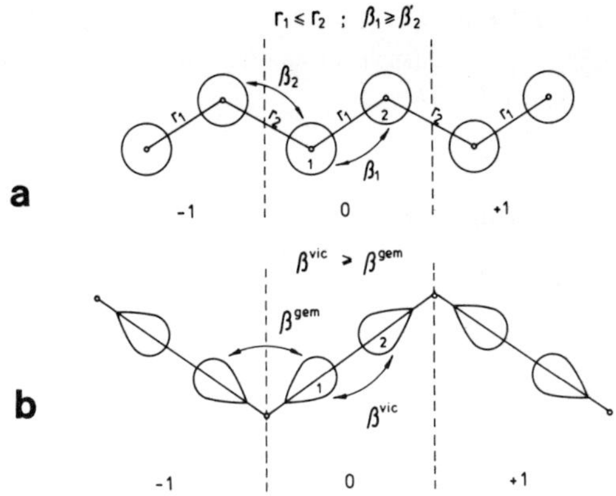

FIGURE 8. The analogy between alternated π and σ electron systems of one-dimensional polymers, the VIFs are identical after $\pi - \sigma$ approximation for polyene and the Sandorfy "C" approximation for polysilane. According to Sinanoğlu's theorem: if a set of molecules or reacting assembles with different structural formulas are related by simple manipulations of their VIFs by the two rules mentioned above, then they have the same level pattern $\{n_+, n_0, n_-\}$.

$$E_1(\mathbf{k}) = \alpha + \sqrt{\beta^2 + \beta'^2 + 2\beta\beta'\cos(\mathbf{k}a)} \tag{27a}$$

$$E_2(\mathbf{k}) = \alpha - \sqrt{\beta^2 + \beta'^2 + 2\beta\beta'\cos(\mathbf{k}a)} \tag{27b}$$

when the nearest-neighbor interaction approximation is used and either $\beta = \beta_1$ and $\beta' = \beta_2$ or $\beta = \beta^{vic}$ and $\beta' = \beta^{gem}$. Clearly, $0 \leq \mathbf{k} < 2\pi/a$ is all that is necessary for nonredundant information.

Using the relations in Equations 27a and 27b one can show that the energy gap:

$$\Delta E \cong 2|\beta - \beta'| \tag{28}$$

will appear on bond alternation (Figure 9).

Peierls and others have shown that the metallic state with β equal to β' in one dimension is inherently unstable and produces a nonzero energy gap connected with the bond length distortions. The whole process can be thought of as a manifestation of the Jahn-Teller effect. For polyene $\beta = \beta_1$ and $\beta' = \beta_2$. Here both integrals are distance dependent. For polysilane $\beta = \beta^{vic}$ and $\beta' = \beta^{gem}$, with β^{vic} being distance dependent while β^{gem} is angle and substituent dependent. As mentioned above, the resonance integral between two sp^3 hybrid orbitals is proportional to $(I_s - I_p)$, where I_s and I_p are the valence state ionization energies of *s* and *p* AOs, respectively. If I_s and I_p were equal, β^{gem} would vanish; however, they are not. The ratio of the resonance integrals between the hybridized orbitals attached to the same atom

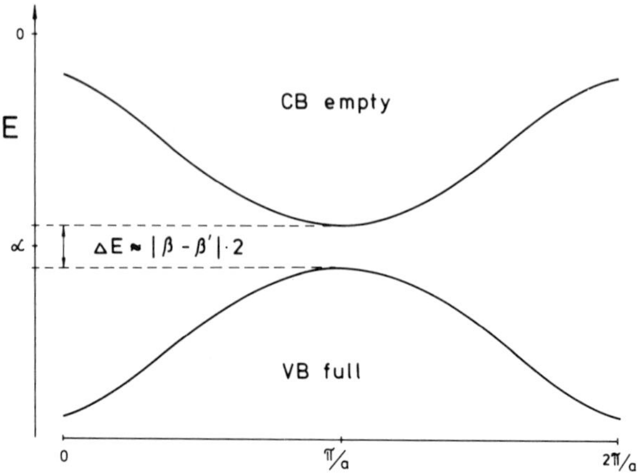

FIGURE 9. Band structure of an alternate one-dimensional infinite semiconductor.

(intra) and between the orbitals that form the Si–Si bond (inter) is m $= \beta^{gem}/\beta^{vic} = \kappa^{-1}$. When m is equal to unity, the electrons are completely delocalized whereas when m is zero, the electrons are localized completely in the σ bonds. The value of m is therefore a measure of the degree of delocalization of the electronic structure.[32]

The parameters β^{gem} and β^{vic} obtained from PES data fitting to simple Hückel-like theory lead to the values of m equal to 0.76 ± 0.01 for alkanes and 0.87 ± 0.01 for polysilanes. The values obtained for m were used to interpret the transmission of substituent effects through carbon and silicon chains; and through-bond interaction effects show, in essence, a smaller degree of delocalization in the carbon chain of alkane, a corresponding faster damping of substituent effects, and a generally less "aromatic-like" character.[32]

Since distortion of any bond angle at the silicon atom (X–Si–Y) changes the hybridization and thus β^{gem} integral, polysilanes offer the possibility of transition from the semiconductor through the inherently unstable metallic state to the hypothetical weakly coupled silylenes state in which $\beta^{gem} > \beta^{vic}$, similar to polyene transition.

The greater than expected reduction of the electron density at silicon in compounds with the Si–X bond (X = P, S, and Cl) has been explained by a conjugation of the lone pair or pairs of X with the σ bond electrons.[33] Similar conclusions have been obtained from CT spectral data of polysilanes when some $SiMe_2$ groups were replaced by isoelectronic divalent sulfur atoms.[34]

Despite symmetry, the qualitative behavior of electrons in P-, S-, or Cl-substituted polysilanes can be predicted in a simple pictorial way from their VIFs using Sinanoğlu's method.

Primary, secondary, or tertiary silylphosphines lead to similar VIF reductions and LPIs (Figure 10a). Thus, phosphorus-substituted polysilanes

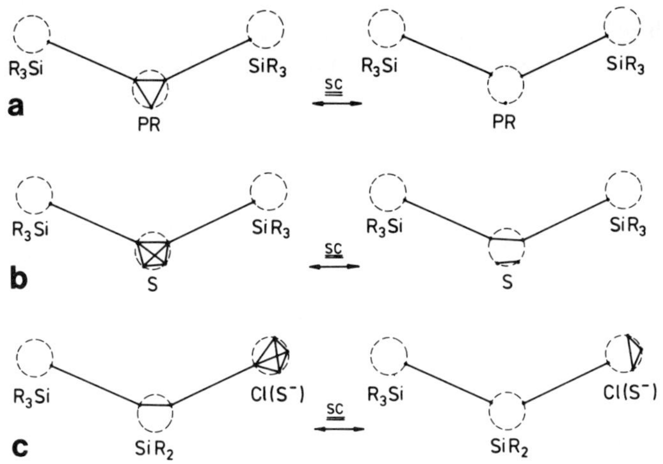

FIGURE 10. VIF of P-, S-, or Cl-substituted polysilanes and its $\overset{SC}{=}$ reductions. VIFs did not show any loops or line strength differences, implying the self-energies and interaction lines of P, S, and Cl do not deviate much from Si where standard values are taken. However, Sinanoğlu's rules show that even if ξs and/or κs are added, the LPI is unaffected.

have LPI = $\{n_+ = n_-$ = number of Si–Si and Si–P bonds and $n_0 = 1$ (at the phosphorus tetra-VP self-energy)$\}$. As in polysilanes, any difference in Si–Si and Si–P line strengths (κs relative to say β_{Si}^{gem}) and in the phosphorus atom VP self-energy ($\xi = [\alpha_P - \alpha_{Si}]/\beta_{Si}^{0}$) do not affect the VIF reductions or the LPI. Thus, in the phosphasubstituted polysilanes too, the numbers of the bonding, nonbonding, and antibonding MOs in the delocalized description of electronic structure can be calculated directly from the structural formula. The VIFs, i.e., the structural electronic formulas, also give the structurally covariant localized descriptions.

In sulfur-substituted polysilanes the sulfur atomic VIF is also tetra-VP. Thus, the VIF of the thio-vicinity in the said silthianes and silanethiols is as in Figure 10b. The fact that the Si–S line strengths may be different from the intrasulfur ones does not affect the LPI. Any sulfur VP loops are also ineffectual. The sulfur vicinity is the same in the above silthianes and silanethiols, the LPI consisting of $\{n_+ = n_-$ = number of Si–Si and Si–S bonds $+2\}$. In the reduced VIFs, the intrasulfur lone VP line gets four electrons in the ECI, leading to two lone pairs as bonding and antibonding ones. The appropriate antibonding orbital has constant energy; and since other sulfur orbitals are lowered by bonding interactions with silicon atoms, the antibonding orbital is unaffected and is a HOMO orbital according to simple Hückel-like calculations, which remains in agreement with experiments.[34]

In chloropolysilanes or polysilanethiolate anions any straight or branched polysilane chain has the same effect after its reduction (Figure 10c) (remarks similar to those above apply also for P, S, or Cl self-energy and line strengths).

The three chlorine "lone VPs" remain triangularly coupled in Figure 10c, contributing one bonding and two antibonding MOs (or local MOs) to the LPI. The triangle accommodates six electrons of the ECI forming three coupled "lone pairs". The full MO LPI is $\{n_+$ = number of Si–Si, Si–Cl (Si–S^-) bonds $+1, n_-$ = number of Si–Si, Si–Cl (Si–S^-) bonds $+2\}$.

At this point one should notice that our discussion was restricted to the Sandorfy "C" approximated orbitals (frontier orbitals). It is now necessary to consider the orbitals used in the bonding of the substituents to the silicon backbone, as in Sandorfy's model "H". The two approximately sp^3 hybrids used by each silicon atom for this purpose combine with the orbitals of substituents into low-energy bonding σ_{SiC} and high-energy antibonding σ^*_{SiC} orbitals. The σ_{SiC} orbitals not depicted in Figure 9 are present in a lower region of the valence band consisting mostly of the σ_{SiSi} orbitals. Similarly, the σ^*_{SiC} orbitals are present in the upper part of the conduction band consisting mostly of the σ^*_{SiSi} orbitals. Symmetry adaptation with respect to mirroring in the plane of silicon atoms yields π symmetry (antisymmetric) bonding and antibonding combinations. Somewhat weak interactions between such orbitals located at neighboring silicon atoms bring about a perfect delocalization along the length of the chain. Since all the Si–Si distances are nearly equal, the overlap and resonance integrals for such delocalization do not alternate in size along the chain. They are larger for the antibonding σ^*_{SiC} orbitals that are largely localized on silicon atoms, whereas the bonding σ_{SiC} orbitals are primarily localized on the substituent. This is caused by the low electronegativity of silicon relative to carbon and would not necessarily be true in otherwise substituted polysilane chains. This also makes both the σ and π symmetry combinations of the σ_{SiC} orbitals somewhat low in energy, well below the less stable among the Si–Si bonding molecular orbitals of the silicon backbone.

The π symmetry combinations of the σ_{SiC} and σ^*_{SiC} orbitals are not permitted by symmetry to mix with any orbitals of the silicon backbone described above, while σ symmetry combinations are. The mixing is of some importance but does not modify very significantly conclusions reached previously for the frontier orbitals of Figure 9.[15]

3. Excited Singlet States of Polysilanes and Related Problems

The information carriers in conventional Si electronics are given by electrons and holes that carry negative and positive charges and spin $^1/_2$. As particles and antiparticles, they emit energy in the annihilation or recombination process. The semiconductor electronics of today are based on these particles that are produced by n- or p-doping of the crystals.

In inorganic semiconductors, the atoms are connected by covalent bonds in all three dimensions. Carrier mobilities are quite high, so that reasonable conductivities are achieved at fairly low doping levels, of the order of several parts per million. Doping is usually substitutional and generally does not lead

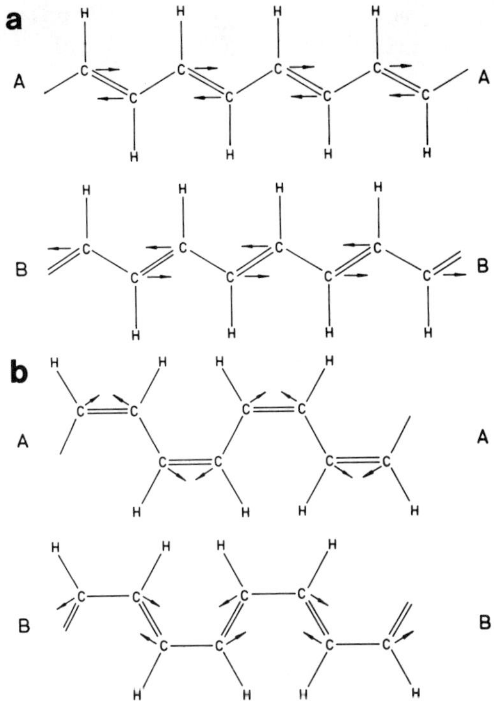

FIGURE 11. (a) Chemical structures of the two degenerate ground-state forms of *trans*-polyacetylene, $E_A = E_B$. (b) Two inequivalent structures for the *cis*-polyacetylene, $E_A < E_B$. The atomic distortions are shown schematically by the arrows.

to major disruption of the lattice. Doping of inorganic semiconductors results in the appearance in the gap of electronic levels that can be thermally populated at room temperature. Rigid-band models give a good description of the physics of these systems.

Conventional polymers are famous for their insulator properties. The systems discussed concerning molecular electronics are conjugated polymers. The simplest systems are shown in Figure 11, and the polysilane system is shown in Figure 12 with the alternative structures A and B.

Structure B is obtained from structure A by flipping of the bonds. The possibility of changing the structure from A to B leads to quasiparticles such as solitons when $E_A = E_B$ and polarons and bipolarons when $E_A \neq E_B$.

In conducting polymers, the interactions are highly anisotropic since atoms are linked by covalent bonds along the polymer chains, whereas interactions between chains are much weaker (usually of the Van der Waals type); this anisotropy opens the door to possible collective instabilities, typical of quasi-one-dimensional materials, such as Peierls distortions. Doping levels are around a few percent, so that some authors use the term loading rather

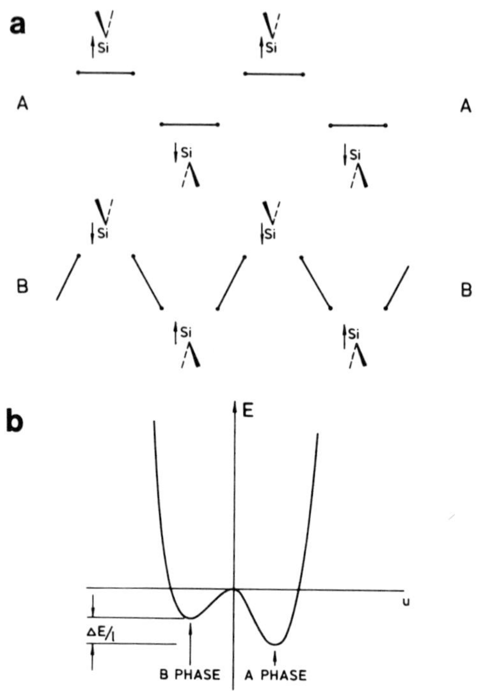

FIGURE 12. (a) Two unequivalent structures for the polysilane as weak (A) and strongly coupled (B) polysilylene, $E_A > E_B$. (b) The two configurations (A phase and B phase) are nearly (but not precisely) degenerate as shown in the diagram, where we plot energy against the distortion parameter, u ($u = 0$ when $\beta = \beta'$). The energy difference per bond, $\Delta E/l$ might be expected to be small, i.e., equal to $2 \mid \beta - \beta' \mid$. The author believes that experimentally observed thermochromism of polysilanes is simply related to $B \rightarrow A$ phase transition that depends on substitution of the backbone silicon atoms, polymer-solvent interactions and chain conformation. For more common interpretation of polysilane thermochromism see Reference 15.

than doping;[62] dopant atoms or molecules intercalate between the polymer chains and undergo charge transfer with the chains. It is expected that charge-transfer (or electron-excitation) processes result in significant local modifications (relaxations) of the chain geometry. Indeed, in conducting polymers, the equilibrium geometry in the ionized state is usually quite different from that in the ground state. In solid-state physics terminology, this means strong electron-phonon coupling. Then, the local geometric modification of the chain markedly affects the electronic structure by inducing localized electronic states into the gap. These levels are all attributable to charge-transfer induced modifications of the polymer and in that sense are intrinsic to the situation in doped inorganic semiconductors where the states in the gap are dopant levels. Because upon ionization or electron excitation, a local relaxation of the lattice geometric and electronic structures is energetically favorable; or in other words

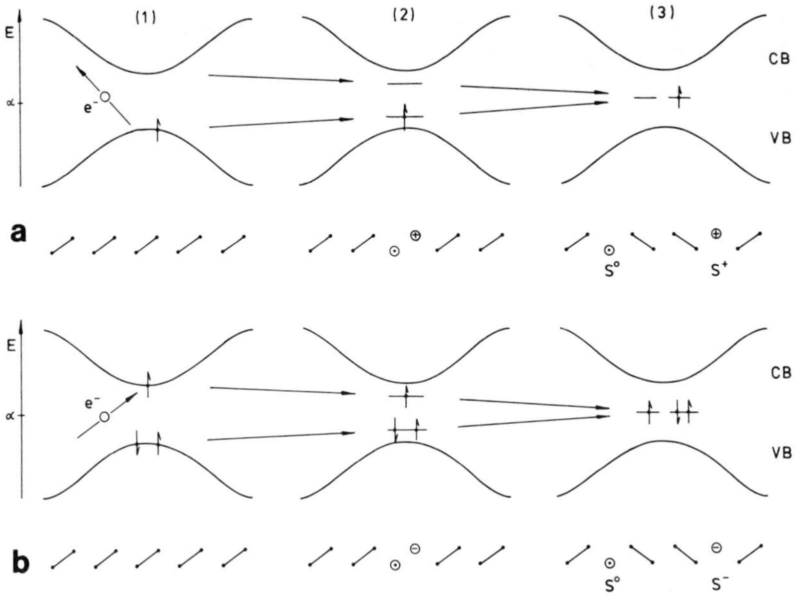

FIGURE 13. Energy dispersion (k space) and VIF (real space) of the quasiparticles in *trans*-polyacetylene after doping. S^0 is a neutral soliton, S^+ and S^- are charged solitons. The following processes are shown: (1) removal of an electron (upper part) and addition of an extra electron (lower part), (2) relaxation to polarons, (3) dissociation into noninteracting solitons.

that charged defects are formed, constitutes the basis of the fascinating physics occurring in doped conjugated polymers.

Here we first discuss the qualitative ideas involved in the formation of charged defects upon doping of *trans*-polyacetylene, a system with a degenerate ground state, and compare the situation in *cis*-polyacetylene and polysilane polymers without a degenerate ground state.

Due to its degenerate ground state, *trans*-polyacetylene represents a special material, which leads to the existence of solitons. In contrast to electrons and holes in conventional semiconductors, charges and spins can be separated in *trans*-polyacetylene. This process is shown in Figure 13 in the energy level diagram and in the VIF, respectively, both in k space and real space.

Let us first take a closer look at the upper part of Figure 13. A hole is created by extracting an electron from the valence band, leaving behind a positive charge and an unpaired electron at the chain. The hole is unstable in a conjugated chain due to lattice relaxation. Due to exchange interaction the charge and spin will stay together and form a polaron. In the dissociation process they will separate and become completely independent particles called solitons.[63] In the separation process two states will split, one from the valence band and another one from the conduction band. Since there is an interaction between the two solitons, the two states will be separated. However, when the distance between the solitons is large, the states will lie exactly centered in the middle of the energy gap. The two particles denoted by S^0 and S^+ can be visualized as a kink or domain wall between the two degenerate forms A and B.

Because of strong electron-phonon coupling, the wave function of these kinks is delocalized over something like 12 to 14 VPs on the average. In the terminology of nonlinear field equations[63] these kinks can be formulated as topological solitons, i.e., nonlinear excitations connecting two degenerate ground states. Their optical and spin properties have been investigated in great detail.[64,65] The lower part of Figure 13 sketches the opposite situation when an electron is added to the chain. Again two solitons result, one with charge $Q = -1$ and spin $S = 0$ and another one with charge $Q = 0$ and $S = {}^1/_2$. It should be noted that a prerequisite for these solitons to occur is a degenerate ground state. Each soliton is accompanied by a large deformation of the polyacetylene chain. The mobility of the soliton is therefore limited to the sound velocity. If the distance the soliton must travel in some molecular circuit is only, for example, 10 nm, a transit time of a few picoseconds can be achieved. As was pointed out *trans*-polyacetylene is an exception because it is the only conjugated polymer that has a degenerate ground state for infinitely long chains. All other conjugated polymers do not have degenerate ground states, and the creation of information carriers is therefore quite different.

For *cis*-polyacetylene there are two VIFs corresponding to local energy minima but they do not have the same energy ($E_A < E_B$) as in the *trans*-isomer when ($E_A = E_B$). The situation is depicted in Figure 11. An obvious consequence of lack of degeneracy is that *cis*-polyacetylene (and similar polymers) cannot support stable kink excitations, since creating a pair of solitons separated by a distance d would cost energy $\sim d(\Delta E/l)$, where $\Delta E/l$ is the energy difference per bond between the stable (A) and metastable (B) phases. Clearly this linear "confinement" of the energy will prohibit solitons from becoming free. Of course, a confined soliton pair is topologically reminiscent of a polaron (Figure 14). Therefore, polarons can be anticipated in *cis*-polyacetylene.

Although the polysilanes are insulators in the pure state, West and coworkers[15,66] have demonstrated that these polymers can be doped to a semiconducting level with arsenic pentafluoride. Now using the VIFs analogy between conjugated π and σ electron systems we will discuss the semiconductivity of polysilanes.

If an electron is extracted from the polysilane chain as shown in Figure 15, a positive charge and spin remain on the chain.

Charge plus spin represent a hole. Now as in *cis*-polyacetylene, the charge and the spin cannot be separated infinitely. They are bound by a constant force, because the energy required for separation is linearly dependent on the

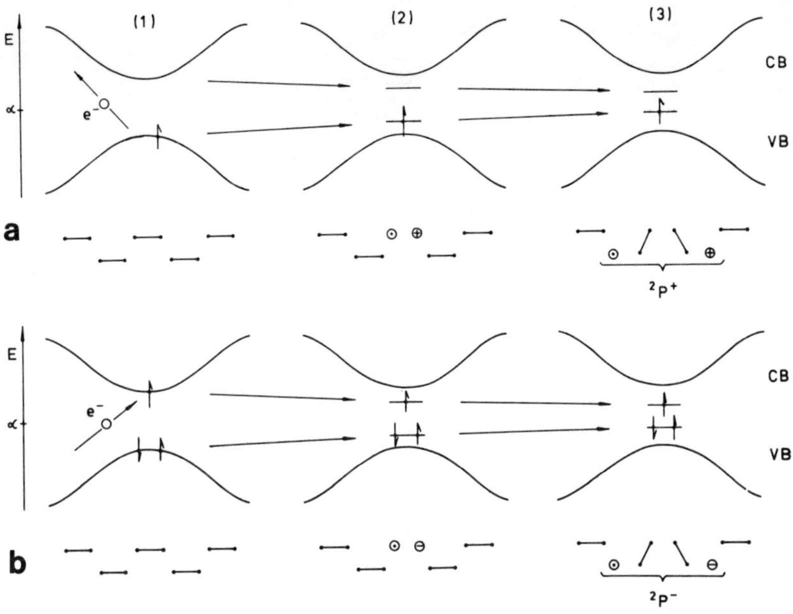

FIGURE 14. Energy dispersion (k space) and VIF (real space) of the quasiparticles in *cis*-polyacetylene after doping. $^2P^+$ and $^2P^-$ are polarons. The following processes are shown: (1) removal of an electron (upper part) and addition of an extra electron (lower part), (2) hole and electron, (3) relaxation to polarons.

distance between the charge and the radical electron. On the other hand, compression of the distance between solitons leads to electronic energy. Therefore, a certain spatial extension of the given polaron P^+ is obtained by the minimum in energy. In contrast to the hole, which is obtained in the first instance, the polaron P^+ is a stable quasiparticle on the polysilane chain. The same arguments hold for the P^- polaron, which can be created by adding an electron to the polysilane chain (lower part of Figure 15).

After this general description of polaron states of doped polysilanes, we will discuss briefly photoexcitations in polysilanes. Again, the VIFs analogy between conjugated π and σ electron systems will be used. A schematic diagram of the photogeneration of a charged soliton-antisoliton pair in *trans*-polyacetylene is shown in Figure 16 (upper part).

The incident photon (for $h\nu \geq \Delta E$) generates an electron-hole pair within the rigid lattice. The system rapidly evolves to a soliton after a time of order 10^{-13} s; the results imply the presence of kinks separating degenerate regions.[64]

The topological degeneracy in *trans*-polyacetylene is not present in *cis*-polyacetylene. Since the A phase (Figure 11b) has a lower energy than the B phase, the solitons (domain walls) would separate nondegenerate regions.

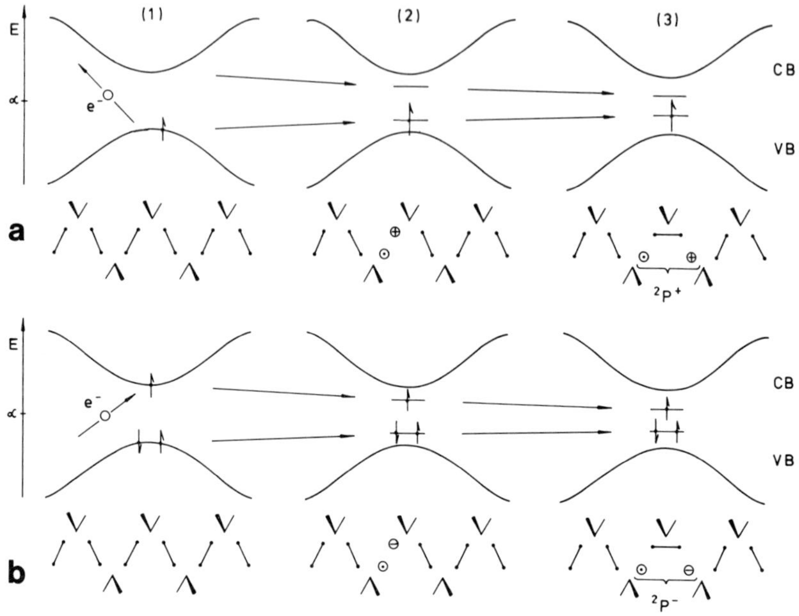

FIGURE 15. Energy dispersion (k space) and VIF (real space) of the quasiparticles in polysilane chain after doping. $^2P^+$ and $^2P^-$ are polarons. The following processes are shown: (1) removal of an electron (upper part) and addition of an extra electron (lower part), (2) hole and electron, (3) relaxation to polarons.

Although neither of the limiting forms A and B is literally correct, we may think of the ground state as essentially equivalent to phase A, with phase B being slightly higher in energy. Let us consider photoexcitation of the charged soliton-antisoliton pair. The energy required for creation is

$$E_{tot} = 2E_s + n/2(\Delta E/l) \tag{29}$$

where E_s is the energy for creation of a single soliton (analogous to the soliton creation energy in *trans*-polyacetylene), n is the number of VPs separating the two kinks, and $\Delta E/l$ is the energy difference per bond between the two configurations A and B. Thus as they begin to form, the two solitons would be ''confined'' or bound into a bipolaron ^{1}BP entity, the further apart, the greater the energy.[64] In contrast to the charged quasiparticles that are intrinsically stable because they can decay only by recombination with their antiparticle, the neutral particle as ^{1}BP is intrinsically unstable. As a result, one expects the photogenerated pair of carriers in *cis*-polyacetylene to quickly recombine. The results are summarized in Table 3.

Photoluminescence is observed from *cis*-polyacetylene, but not from *trans*-polyacetylene, whereas photoconductivity is observed in *trans*-polyacetylene, but not in *cis*-polyacetylene.

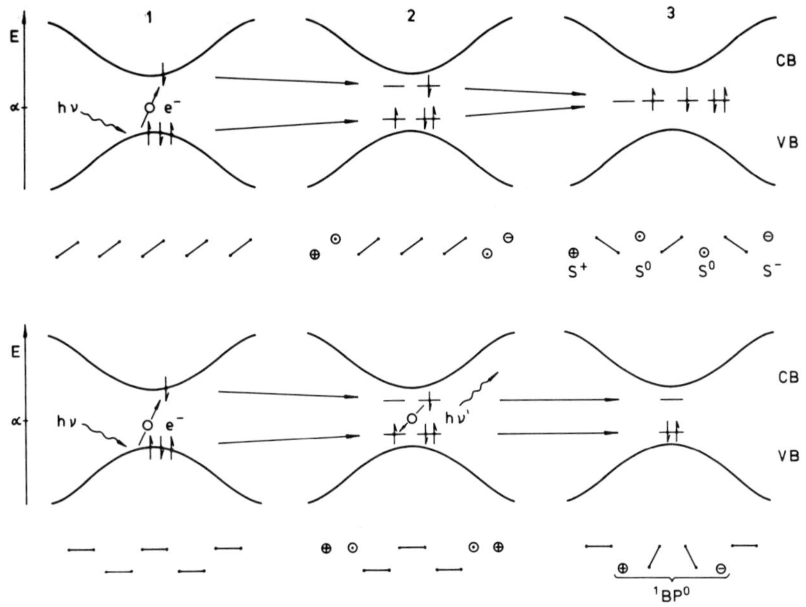

FIGURE 16. Energy dispersion (k space) and VIF (real space) of the excited *trans*-(upper part) and *cis*-(lower part)polyacetylene. S^+, S^0, and S^- are solitons and ^1BP is a neutral bipolaron. The following processes are shown: (1) transfer of an electron by incident photon (for $h\nu \geq \Delta E$), (2) generation of an electron-hole pair, (3) rapid relaxation to charged soliton-antisoliton pair. The solitons are free and can move apart with no cost in energy (upper part) or the solitons are confined; the farther apart the greater the energy (lower part), thus the creation of a neutral bipolaron is favored.

TABLE 3
Summary of Principal Polyacetylene Properties

Property	*cis*-$(CH)_n$	*trans*-$(CH)_n$
Ground state	Nondegenerated	Degenerated
Nonlinear excitation	Confined solitons	Free solitons
Luminescence	Yes	No
Photoconductivity	No	Yes

Data from Reference 64.

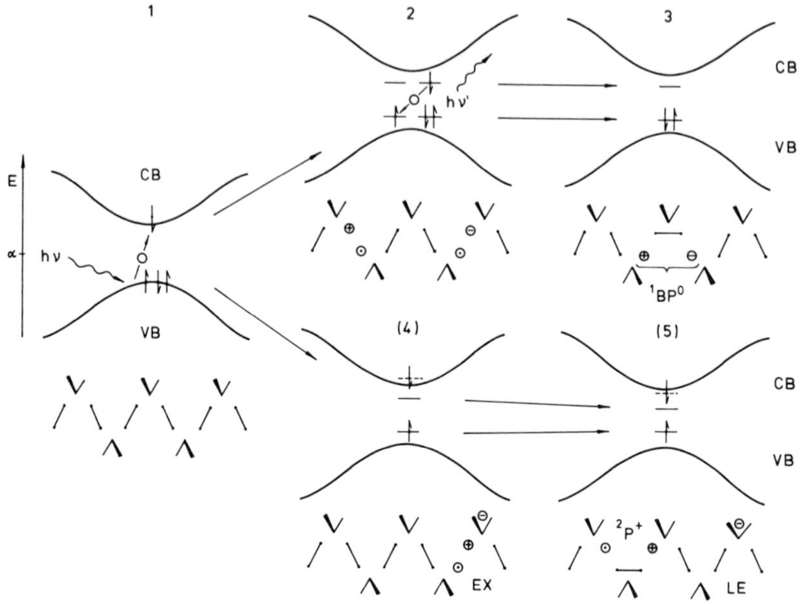

FIGURE 17. Energy dispersion (k space) and VIF (real space) of the excited polysilane (^1BP, bipolaron; EX, created exciton; $^2P^+$, polaron, and LE, localized electron). The following processes, are shown: (upper part) canal analogous to *cis*-polyacetylene, (lower part) canal specific for polysilane. (1) transfer of an electron by incident photon (for $h\nu \geq \Delta E$), (2) generation of an electron-hole pair, (3) the creation of a neutral bipolaron, (4) exciton state created directly or converted internally from (1), (5) localized electron and free positively charged polaron (the localized π^*_{SiR} orbitals are marked as dashed).

Polysilanes possess characteristic molecular weight dependent, strong unstructured fluorescence and excellent photoconductivity, which bear resemblance to those of *cis*- and *trans*-polyacetylene. This unusual mix of properties suggests more complicated photocarrier generation in polysilane than in *cis*-polyacetylene. The explanation of these properties is shown schematically in Figure 17.

In the first canal (upper part) an electron-hole pair is generated after $\sigma_{SiSi} \rightarrow \sigma^*_{SiSi}$ excitation. Relaxation of the polymer chain leads to the formation of two polarons $^2P^+$ and $^2P^-$; decay into $^1BP^0$ state causes the emission shown in Figure 17.

The second canal (lower part) begins with formation of an excimeric state by initial $\sigma_{SiSi} \rightarrow \pi^*_{SiR}$ transition or by nonradiative transition $\sigma^*_{SiSi} \rightarrow \pi^*_{SiR}$ (internal conversion). The two types of valence transitions, $\sigma_{SiSi} \rightarrow \sigma^*_{SiSi}$ and $\sigma_{SiSi} \rightarrow \pi^*_{SiR}$, compete nearly evenly for the position of lowest energy, and the outcome of the competition is a function of chain length. The $\sigma_{SiSi} \rightarrow \sigma^*_{SiSi}$ excitation energy decreases more rapidly than the $\sigma_{SiSi} \rightarrow \pi^*_{SiR}$ excitation energy as the number of backbone silicon atoms increases.[68] Since the σ_{SiSi}

$\rightarrow \sigma^*_{SiSi}$ excitations obey the selection rules for electric dipole radiation, they are permitted ones whereas all the $\sigma_{SiSi} \rightarrow \pi^*_{SiR}$ transitions are electronically forbidden. Relaxation in this canal leads to long-lived π^*_{SiR} localized electron LE and free positively charged polaron $^2P^+$ states.

Workers at Sandia National Laboratories and Nippon Telegraph and Telephone have studied a variety of polysilane derivatives and found them to be excellent photoconductors.[15] In each study only positive carriers were found mobile, and nondispersive transport was observed with measured mobilities of order 10^{-4} cm^2/(Vs) at room temperature were reported. The insensitivity of the carrier mobilities measured at room temperature to the nature of the polymer substituents lead to the conclusion that transport occurs via the σ electronic states of the silicon backbone rather than by hopping between side-chain substituents. The measured positive carrier mobilities depend on the applied electric field and are sensitive to temperature. The drift mobility is thermally activated, and activation energies of 0.25 to 0.28 eV have been measured. The measured activation energies are among the lowest values reported for amorphous solids, and the drift mobilities are quite high. Kepler et al.[69] studying photoconductivity of poly-(methylphenylsilane) have estimated that the photocarrier generation quantum efficiency approaches 1% at high electric fields and that the carrier lifetimes are in excess of several milliseconds. Carrier lifetimes are affected by impurities and decrease mostly in the presence of certain electron-donor dopants.

In traditional "organic" one-dimensional π semiconductors, the σ bond framework compressibility leads to a more rigid structure than in polysilanes. The situation in polysilanes is different; charge carriers localized mainly in σ framework makes these systems generally more susceptible to structural distortion (for example, the Si–Si bond force constants are between 1.65 and $2.3 \cdot 10^{-5}$ dyn/cm[70]). As a result, the dominant charge carriers are inherently coupled to chain distortions. This will tend to localize the optical excitations.

At this stage of the argumentation, it is thus necessary to compare in a more detailed way the electronic excitations in the polyene series (whose last term is polyacetylene) with transitions in the polysilane oligomers; their relations to the polymeric band gap; and the role of correlation on excitations, localization of excitations, and size of band gaps. The expected predominant role of exchange interactions can manifest itself as a certain localization of excitations.

We will discuss these topics taking the SCF CI(1) VPs interaction framework as a reference. Results of the calculations at this level of approximation revealed that near UV transition bands are almost quantitatively predicted from the point of view of energy and intensity.[36] Also the calculated ionization potentials are predicted very well, with an average deviation of 0.12 eV.

Basic ideas can be introduced by examining the dependence of the first UV transition on the size of the polyene and polysilane chain. Figure 18 depicts the results of SCF CI(1) σ electron and π electron calculations. The

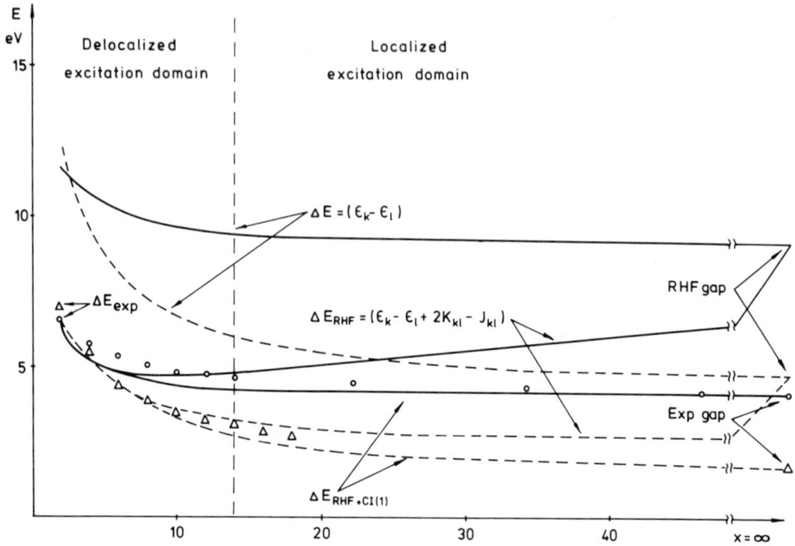

FIGURE 18. SCF CI(1) σ electron calculations on polysilane chains (circles for experimental values and solid lines for calculated one) and π electron calculations on polyene chains (triangles for experimental values and broken lines for calculated values).[71] Evolution of the HOMO-LUMO separation $\Delta E = \epsilon_k - \epsilon_l$, the first singlet RHF transition energy ΔE_{RHF}, and the first singlet CI(1) transition energy ΔE_{RHF+CI} with respect to the size of the chain in number of VPs. The minimum on each RHF transition energy curve separates a domain where one-configuration excitation dominates for small oligomers from a domain where many configuration excitations must be considered for large oligomers.

property of interest is the evolution of the first UV transition with the size of the polymeric chain. The first UV transition here is decomposed into its three constituent terms (lowest unoccupied molecular orbital [LUMO] and HOMO orbitals energy difference, transition, Coulomb and exchange contributions, as well as CI(1) effects) and compared to available experimental data.

Due to a decrease of Coulomb and exchange contributions accompanying an increase of the number of VPs in the chain, the resulting quantity, which is a theoretical measure of the fully delocalized one-configuration singlet-singlet transition energy, reveals a minimum and then slowly rises to $\approx$9 eV (polysilane) or $\approx$5 eV (polyene) for an infinite number of VPs. The experimental curve does not exhibit such a minimum and asymptotically approaches the band gap at $\approx$4 eV for polysilane or $\approx$2 eV for polyene.

For both polymers, the restricted HF (RHF) energy gap is much larger than the experimental and assumed extrapolated one. This is a known artifact of the one-configuration SCF model. Concerning this point, a careful analysis of Figure 18 calls for the following comment. The agreement between one-configuration theory and experiment is good for chains of small size (less than $\approx$10 VPs for polysilane or $\approx$30 VPs for polyene). This delimits the frontier where delocalized one-configuration excitations are no longer significant for explaining electronic transitions and where many-configuration localized excitations should be explicitly taken into account. It can be inferred from the figure that the extent of localized many configuration excitations is somewhere between 8 or 10 VPs for polysilane and over 20 or 30 VPs for polyene and that the difference between the RHF and RHF + CI excitation energy is a rough measure of the energy difference between the fully delocalized one-configuration and localized many-configuration excitations. It is tempting to relate those results with considerations at the Sinanoğlu level. These considerations indeed show stronger bond conjugation and the stronger lattice deformation due to the presence of excitation in polyacetylene than in polysilane.

This idea is supported by the following arguments. The transition Coulomb and exchange correction terms ($J_{kl} - 2K_{kl}$) to the SCF orbital energy difference is needed in order for the electron promoted in a virtual orbital to properly feel the influence of ($N - 1$) electrons potential rather than N electrons potential, N being the total number of electrons. For large systems, this transition correction term vanishes for fully delocalized orbitals as the modification of Coulomb and exchange interactions due to the promotion of a single electron of a large N electron system is negligible. In physical reality, there most likely exists an important electron-hole correlation in considered conjugated polymers. As a result, it can be expected that in the larger molecules, the excited electron is actually localized around the hole and not delocalized all over the molecule. A theoretical confirmation of this fact is given by CI calculations including single excitation corrections in the first excited state. The spectacular first order effect of this CI calculation is that it completely removes the artificial RHF minimum of the transition curve with the size of the chains and produces a curve in good agreement with experimentally observed changes. The localization effect is depicted by the analysis of the changes in VP charges (Δq) between the first singly excited and the ground state. RHF Δq values are very small (of the order of 10^{-4} electron) and therefore show that the VP charges remain about the same when an electron is promoted from the HOMO to the LUMO. The reason for this is basically that the linear combination of atomic orbitals (LCAO) coefficients of the HOMO and LUMO are almost identical in absolute values since these two orbitals are nearly symmetrical with respect to the reference state. A quantitative estimate is that for a chain of 40 VPs of polyene, the charge gain on the 16 inner VP is of the order of 0.36 electron,[71] similar effects are in polysilanes. This clearly supports an interpretation in terms of localization of the excitation in the middle of the molecule. This localization effect also can be viewed as an exciton or a correlated electron-hole pair formation.

Linear combinations of VPs models provides a complete basis for clarifying differences and similarities between σ- and π-conjugated polymers.[72,73]

The one-photon excitation to 1^1B two-photon absorption to 2^1A_g, and lowest triplet excitation 1^3B are properly predicted for permethylpolysilane oligomers and polymers.

The two-photon absorption 2^1A is about 1.0 eV above the one-photon absorption for these calculations in agreement with experiment. The SCF CI(1) model for polyenes predicts that 2^1A_g is below 1^1B_u and both agree with available data up to N = 12 polyene VPs. The reversal of one- and two-photon excitations in polysilane is primarily an alternation effect in the model, whose intersite Coulomb interactions produce an excitonic state at 1^1B.

The position and intensity of two-photon absorptions are both sensitive to electron-electron correlations. The electron-hole symmetry of regular infinite VP chains leads to a vanishing excitation energy for 2^1A_g, while the strong one-photon absorption at the optical gap remains finite. Since two-photon intensities are related to $\mathrm{Im}\chi^{(3)}(-\omega,\omega,-\omega,\omega)$, they may be found exactly from two-photon transition moments. The oscillator strength per site is essentially independent of N in this model in agreement with experiment for both polyenes and polysilanes. Higher order responses, by contrast, increase rapidly with oligomer size N up to some critical value, N_c, corresponding to the conjugation length. Polysilane absorption and fluorescence data are widely discussed in terms of short ordered segments of VPs in far longer chains. The two-photon intensity per site is sensitive to chain conformation, while the one-photon intensity is almost identical.[72]

Quantum chemistry is now providing efficient ideas for the molecular design of materials with highly specific optoelectronic characteristics.[74] In this respect, σ-conjugated polysilane polymers have recently gained an important technological interest.[75]

The σ-conjugated polysilane polymers seem to be suitable candidates for materials of high efficiency in nonlinear optics, mainly for third order polarizability related properties and third harmonic generation (THG). For THG properties the idea of an effective chain length or of an electron delocalization length is crucial.[76]

A good test of the applicability of the model is an attempt to extend it to other kinds of polysilanes. A comparison of the calculated and observed singlet electronic transitions for the dialkylpolysilylenes $(R_2Si)_n$ is given in Reference 36. The disilene molecule (n = 2) will be discussed here as an interesting example. The SCF-MO energy levels considered in CI(1) calculations are illustrated in Figure 19.

An interesting feature found in the energy level diagrams is that the LUMO of disilene has σ* symmetry, whereas that of ethylene has π* symmetry. The calculated and experimental transition energies and moments are shown in Table 4.

Agreement between the calculated and available experimental properties of tetramethyldisilene is satisfactory; all transitions in typical UV regions are forbidden, and the first transition energy of disilene is perfectly predicted. Since the σ* (−1)-orbital and the π (1)-orbital possess different symmetry, they cannot interact (transition moment is zero). Interaction of both orbitals

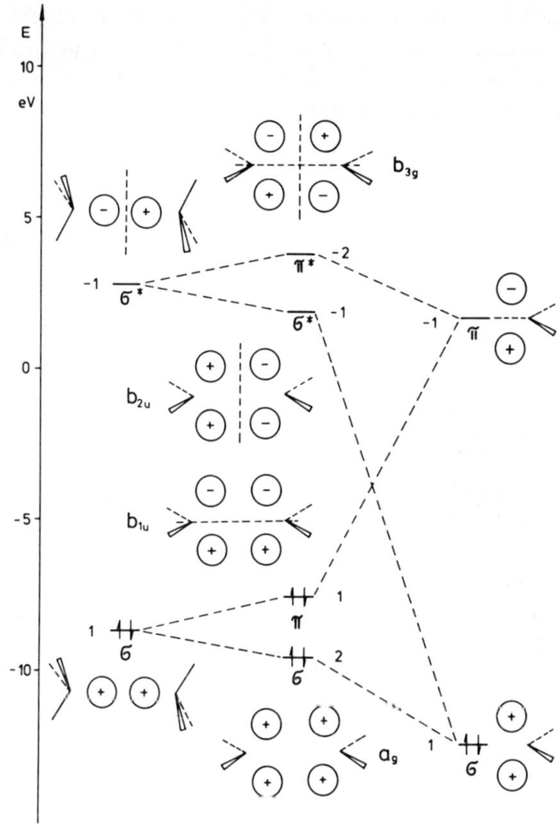

FIGURE 19. MO energy level diagrams for disilane (*1*), disilene (*2*), and silylene (*3*).

TABLE 4
Comparison of Calculated[36] and Observed Singlet Electronic Transitions[77] for Hexamethyldisilane (1), Tetramethyldisilene (2), and Dimethylsilylene (3)

	Calculated			Experimental	
Compounds	**ΔE**	***f***	**Φ_J^{-K}**	**ΔE^a**	**$\log_{10} \epsilon$**
1	6.41	0.488	$1.000\ \Phi_1^{-1}$	6.41	3.90
2	3.10	0.000	$1.000\ \Phi_1^{-1}$	3.10	—
	4.97	0.000	$0.707\ \Phi_2^{-1} + 0.707\ \Phi_1^{-2}$		
	6.84	0.000	$1.000\ \Phi_2^{-2}$		
3	2.75	0.073	$1.000\ \Phi_1^{-1}$	2.75	3.18

[a] ΔE values in eV.

can only occur via symmetry breaking. The mixing of π orbital with the σ^* one requires a vibration according to the direct product of the irreducible representations of the two orbitals (Equation 30). The b_{3g} distortion is caused by the introduction of antipyramidization on the Si atoms:[78]

$$\Gamma_\pi \otimes \Gamma_{\sigma^*} = b_{1u} \otimes b_{2u} = b_{3g} \tag{30}$$

Despite the built-in bias against the banana bond idea, the generalized valence bond method does predict them to occur in Si_2H_4.[44,47] On the other hand, Sinanoğlu's "blackboard mnemonic" method explains the difference between ethylene and disilene ($H_2X = XH_2$, X = C or Si) in a very simple way. First, ignoring the hydrogen and X atom valency point energy difference, the VIF is as shown in Figure 20. Taking an H tip onto X atom tips on the same atom of the molecule, the rules immediately reduce (I) to (II), the alternating tetragon (its LP is $n_+ = n_- = 2$) and four separate X-H lines (each X-H gives an additional $n_+ = n_- = 1$ as seen in Figure 20). If $\beta_1 = \beta$, the LP of the tetragon fragment is $n_+ = n_- = 1$, $n_0 = 2$; such distortion leads to a qualitatively less stable structure (VIII) according to Sinanoğlu's rules. Some distortions that preserve the LP ($n_+ = n_- = 2$) are (III), (IV), (V), (VI), and (VII); other distortions change the LP. Figure 20 shows the way of stabilization for the $H_2X = XH_2$ system.

The ratio between vicinal and geminal interactions is decisive. For X = C:

$$\beta_1^{vic} = \langle \chi_p | \hat{h} | \chi'_p \rangle < \beta_2^{vic} = \langle \chi_{tr} | \hat{h} | \chi'_{tr} \rangle$$

and

$$\beta^{gem} = \langle \chi_p | \hat{h} | \chi'_{tr} \rangle = 0 \ (\text{since} \langle \chi_p | \chi_{tr} \rangle = 0)$$

are preferred as in configuration (VII); however, when X = Si we have:

$$\beta_1^{vic} = \langle \chi_{te} | \hat{h} | \chi'_{te} \rangle$$

and

$$\beta^{gem} = \langle \chi_{te} | \hat{h} | \chi'_{te} \rangle$$

with β^{vic} comparable to β^{gem}. The difference between β^{gem}/β^{vic} ratio of carbon and silicon atoms have been explained as follows.[78] In silicon, the valence 3s orbital is more strongly contracted than its 3p valence orbital in opposition to carbon 2s and 2p orbitals that possess comparable exponents of a Slater-type wave function. This trend increases from Si to Sn. Thus, the valence s

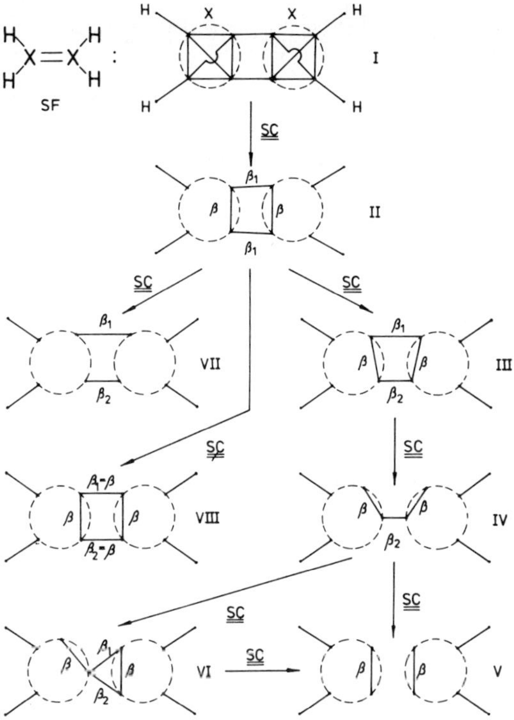

FIGURE 20. Some VIFs of the $H_2X = XH_2$ molecule (X–C or Si). The ($\underset{=}{SC}$) lines between configurations indicate the latter are "structurally covariant", therefore having the same MO level patterns. If ($\underset{\neq}{SC}$), the LPIs differ from the ground-state configuration.

and p orbitals differ in space extension and hybrid atomic orbitals containing large s orbital share (e.g., vicinal $\langle \chi_{tr}|\chi'_{tr}\rangle$) overlap less to form strong interaction $\langle \chi_{tr}|\chi'_{tr}\rangle$. Then configurations (II) to (VI) are preferred as measured by the antipyramidalization tilting angle, which increases from Si to Sn.[78]

Our results obtained from the LCVP SCF CI(1) approach for the polysilanes show that wave functions consisting of equivalent hybrid orbitals of the silicon atom give very good bond description for both polysilanes and polysilylenes.

4. Remarks about Degradation of Polysilanes

α,ω-Dichloropolysilanes are ideal substrates for the synthesis of higher ring and chain polysilanes[79] and polysilane derivatives, because the silicon-chlorine bond in these compounds can react in both coupling and substitution reactions.

Chlorination of cyclic polysilanes gives the best route of preparation of α,ω dichloropolysilanes:

$$(R_2Si)_n \xrightarrow{Cl_2} Cl\text{–}(R_2Si)_k\text{–}Cl, \quad k \leq n \tag{31}$$

The cyclic starting materials can now be routinely prepared with over 90% yield.

The frontier orbital densities calculated at simple Hückel-like levels were successfully used to account for the features of the skeletal cleavage reactions of polysilanes.[30] Comparison of experimental kinetic data for the reactions of oligosilanes with chlorine, with the suitable ionization potentials, suggests that the first step of the chlorine cleavage of polysilanes is a one electron-transfer process from HOMO of polysilanes to LUMO of the chlorine molecule. The same principles apply when other electrophiles are considered.

The photochemical silicon-silicon bonds scission processes of polysilanes are potential uses of these polymers.[80] Polysilanes represent the first examples of stable σ-conjugated high polymers. Their spectroscopic and electronic properties are strongly reminiscent of those of the π-conjugated polymers; however, unlike the π-conjugated polymers, polysilanes are generally fully soluble, processible, and air-stable. Their photosensitivity endows some polysilanes with promise as photoresist materials.[81] These materials are radiation sensitive and are degraded to lower molecular weight fragments upon exposure to UV light and ionizing radiation. For an overview of the nature of these radiation-induced processes and some potential applications, primarily in the field of microlithography, the reader is referred to Reference 82.

Although the simple SCF CI(1) semiempirical model above can predict the IPs and transition energies of the polysilanes remarkably well, there are some uncertainties about the exact nature of the photochemical processes that are involved. The lower electronic transitions of the polysilanes can be unambiguously assigned as $\sigma_{SiSi} \rightarrow \sigma^*_{SiSi}$ transitions because of a very good correspondence between the values computed at the SCF CI(1) level and experimental data[36] and because of the fact that elimination of R_2Si: from polysilane chains is photochemically allowed according to the simple hybrid orbitals interaction approach.[83] Also, evidence has been found suggesting that the silicon-silicon bond order changes calculated at the SCF CI(1) level of the above model can be correlated with the observed experimental ability to extrude a silylene from polysilane chains or rings.[37] For small linear oligosilanes the difference between ground and excited state bond orders appears to be extremely large, because of the small number of delocalized σ_{SiSi} bonding electrons. The longer silicon catenates are more stable and poorer photochemical silylene sources than oligosilanes. One also may expect that lack of equivalence of bond order changes in phenyl-substituted polysilane polymers is responsible for more complicated processes of their photolysis than that of alkyl-substituted polysilanes, with both photofission and photo-cross-linking processes reported to occur.[38]

These facts considered together suggest strongly that polysilanes with unusual optical and electrical properties are predicted to have a high market potential in the future due to high chemical stability with possibility for structure modifications using radiative treatment.

The great wealth of photochemical data on the polysilanes does provide several anomalous situations, e.g., the $\sigma_{SiSi} \rightarrow \sigma^*_{SiSi}$ bands which still lack explanation or observation that the formation of dialkylsilylenes R_2Si: occurs only at somewhat short excitation wavelengths, whereas the photodegradation of molecular weight occurs at all wavelengths that are absorbed.[80]

III. THE COMING OF AGE: PERSPECTIVE

One goal of molecular electronics is to supply computational elements that exploit the molecular rather than bulk properties of materials.[84] The use of molecular properties should allow the design of devices from small to large and the tuning of their properties with molecular modification. In spite of the fascinating variety of molecules that have been proposed as molecular devices, insufficient emphasis has been placed on molecules that are synthetically tractable and could exploit molecular sizes using known chemistry. Synthetic tractability means that the molecules could be assembled using known synthetic techniques. They must derive their function from well-known chemistry, allowing them to perform in a manner that can be modified by making routine synthetic changes to the molecules. To exploit molecular sizes, relatively few large size (micrometer or larger) elements should be needed to operate the device per molecular functional element.

Five challenges must be addressed in the design of a molecular electronic device:[85] delivery of the energy at the molecular level, delivery of the clock signal to the device, interconnection of the molecular devices, communication between the macroscopic word and the molecular device, and dealing with errors.

Molecule-sized devices are appealing because of their potentially large device density and low energy consumption. Many nanometer-sized molecular features are needed per larger (micron-sized) interconnections with the macroscopic device for avoiding error correction problems by operating many molecular devices in parallel. Molecule scale information could be written or processed in a few ways. Molecular conformations, bonding, charge distributions, or oxidation states, for example, could be used as information carriers.

In semiconductors the doping by impurities introduces a micrometer scale asymmetry that is responsible for the operation of all diodelike devices. By analogy an inter- or intramolecular charge transfer between the donor and acceptor ends of a conjugated molecule introduces a nanometer-scaled asymmetry already used in optical parametric amplifiers and soon to be extended to other functions.

A. AN ELECTRON TRANSFER DEVICE BASED ON POLYSILANES

If digital computation is to be carried out at a molecular level, a "bit" must have a physical representation. As already discussed, possibilities include representing a bit by the presence or absence of an electron, an electronic excited state, an exciton, a particular molecular conformation, a spin, etc. The choice made here of the electron (or hole) as the bit has the advantage that it interfaces in a natural way with reading and writing by electrical circuits. In addition, the conservation of electrons keeps bits from being spontaneously created or destroyed. There are also well-developed methods to move electron bits. The absence of a well-developed "copy" procedure is a weakness shared by all the other possible bit representations and warrants conceptual and experimental study.[86]

Because of the ease of writing, shifting, and detecting electrons, particularly with the assistance of light, we have focused on a molecular device based on electron transfer.

Electron transport between two localized electronic states can be controlled by modifying the electronic coupling (tunneling matrix element), the nature of the nuclear modes coupled to the two electronic states (Franck-Condon factor), and the coupling between electronic and nuclear motion.[87]

1. The Polysilanes as "Molecular" Wires

The analysis of this chapter is based on a physical hybrid, consisting of a silicon VLSI-style circuit on which the molecular devices are installed. The silicon circuit will provide the means of making electronic and logical contact with the molecular structures as proposed by Hopfield et al.[86] The base silicon chip would have various metal and oxide parts exposed and be made with 1-μm-scale lithography. Appropriate surface treatments with cleverly designed chemicals and/or electrochemistry will then generate a surface with bonded hydrogen atoms (Si–H or O–H) on which the added polysilane based devices are built by modular silylene insertion into Si–H bonds[56] as indicated in Figure 21.

Short-range self-wiring with polysilanes might replace certain metalization or polysilicon layers in a conventional VLSI circuit. The terminal (Si–H) bond might be simply used for specific polysilane-to-silicon or polysilane-to-metal end binding through dehydrocoupling bond formation.[88] To deal with inevitable errors, assuming a small fraction of the chains have the wrong length or other defects, a few thousands of chains are assembled to discharge a small VLSI floating gate in a few milliseconds.

The polysilanes have one unique property in comparison with other conducting polymers. They are thermally and oxidatively stable yet photochemically labile. This combination of characteristics is ideal for fabrication of high quality complex connections on the VLSI chip since excess of polymer can be removed during exposition to a high intensity laser beam, when the polymer spontaneously degrades and volatizes.

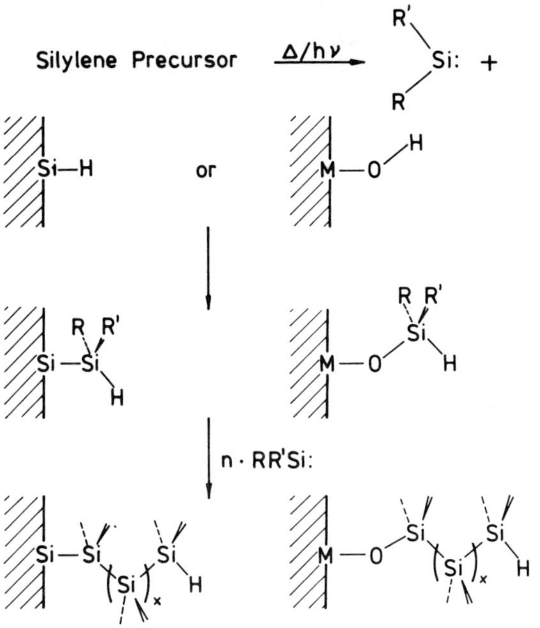

FIGURE 21. Communication between the VLSI circuit and the molecular world might be achieved via polysilane wires of $(RR'Si)_x$ synthesized on chemically prepared silicon or metal surface.

2. Substituted Polysilane Fragments as Potential Electron Tunnel Switches

Moiety or molecular addressability can be achieved by inserting switches into the polysilane chains. Electron tunnel switches stem from work of Pschenichov and Carter[84] and depend on resonant tunneling through series of identical periodic barriers as shown on Figure 22.

Tunneling occurs when the incoming electron energy matches exactly that of a pseudostationary level of periodic wells centered on silicon atoms. The potential distribution and exact levels of pseudostationary states in each well is determined by substituents on the silicon atom. Highly electropositive substituents with empty, low-lying orbitals are required to establish promoted 3 p orbitals of the silicon atom. The same effect is observed when the R–Si–R′ bond angle is increased by repulsion between substituents. Substituents more electronegative than Si preferentially stabilize the 3 s character of the silicon atom. The control donor (D) and acceptor (A) groups on some silicon atoms in the chain can produce changes in hybridization of the silicon atom which mean changes in pseudostationary levels of valence electrons (different s-p mixing), or in other words the local B $\rightarrow$ A phase transition (see Figure 12). By changing the electronegativity of substituents due to ionization the resonant

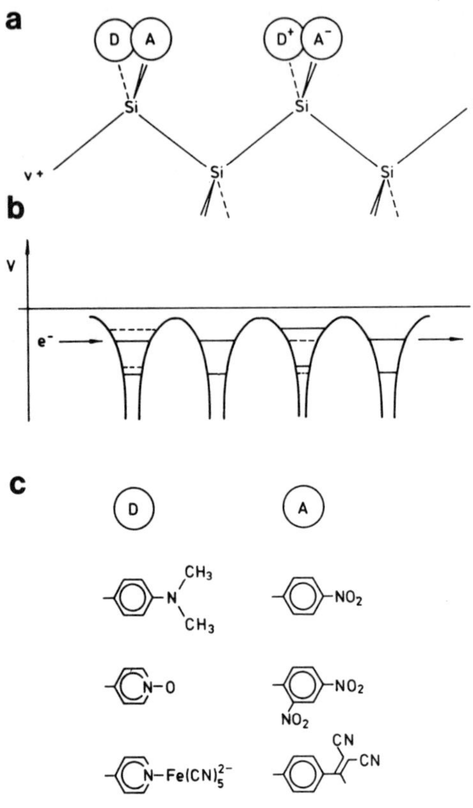

FIGURE 22. A polysilane electron tunnel switch with two dummy control groups D and A. The switch off is done by CT transition between donor and acceptor groups on the controlling silicon atoms bonded with D and A. In part c some examples of donor and acceptor substituents are shown.

tunneling can be sharply switched off since the associate pseudostationary state would move away from the resonant condition due to change in Si 3s and 3p orbitals share. The controlling CT transition as in Figure 22a can be obtained by photostimulation. By modification of the D and A group switching can be made sensitive to wavelength as proposed by Carter.[84] Thus, Figure 22 represents an optically controlled polysilane analog of a NAND gate. Before discussing ideas of information storage we should note that the variety of possible control groups for proposed tunnel switches is large, and that in Figure 22c only first estimations for such groups are presented. Besides charge effects the conformation of aryl-substituent vs. Si–Si bond is also an important factor.[38]

3. Concept of Information Storage

Here a route to higher information density will be discussed. The key feature of molecules that makes them favorable for information storage is the strongest localization of electrons on the molecular skeleton in comparison to bulk silicon. This allows (at least in principle) the use of a cluster of, e.g., 1000 polysilane chains each bonded with a bit storage molecule at the end, as an information carrier. It is obvious that a storage capacity of, e.g., 10^9 b/mm^2 is attainable. In the following we will want to discuss a few basic aspects of this problem.

An approach relies on the attachment of light switchable acceptor end groups (AT) to the polysilane chain through an appropriate spacer using the terminal Si–H bond ($-(R_2Si)_x-(H_2C)_y-AT$). The strategy for achieving this depends on our ability to construct the spacer AT end group and then bond it to the polysilane chain. The most promising are two methods.

The Si–H bond reaction with Cl_2 or CCl_4 produces a Si–Cl bond, which may be converted to an alkoxy–Si bond by reaction with an alcohol:

$$-(R_2Si)_x-H \xrightarrow{Cl_2,CCl_4} -(R_2Si)_x-Cl \tag{32}$$

$$-(R_2Si)_x-Cl + H_3COH \rightarrow -(R_2Si)_x-OCH_3 \tag{33}$$

The –Cl or $-OCH_3$ substituent derived from the Si–H bond can be converted to a Si–C bond by active organometallic compounds such as Grignard reagents or organolithium compounds:[89]

$$\begin{gathered} -(R_2Si)_x-Cl + BrMg(CH_2)_y-AT \rightarrow \\ -(R_2Si)_x-(CH_2)_y-AT + MgBrCl \end{gathered} \tag{34}$$

$$\begin{gathered} -(R_2Si)_x-OCH_3 + BrMg(CH_2)_y-AT \rightarrow \\ -(R_2Si)-(CH_2)_y-AT + MgBrOCH_3 \end{gathered} \tag{35}$$

The silicon-carbon bond can often be conveniently formed also directly by the addition of the Si–H bond to an unsaturated C=C double bond. This can be carried out as a free-radical chain reaction, or more conveniently with catalysis by platinum or other transition metals:[90,91]

$$\begin{gathered} -(R_2Si)_x-H + H_2C{=}CH-(CH_2)_y-AT \xrightarrow{RO\cdot} \\ -(R_2Si)_x-CH(CH_3)-(CH_2)_y-AT \end{gathered} \tag{36}$$

$$\begin{gathered} -(R_2Si)_x-H + H_2C{=}CH-(CH_2)_y-AT \xrightarrow{H_2PtCl_6} \\ -(R_2Si)_x-(CH_2)_{y+2}-AT \end{gathered} \tag{37}$$

As the equations show, we really have a free space for constructing optimal synthesis strategy.

The acceptor may be excited by light resulting in an acceptor excited state that permits electron transfer via the bridge-spacer to the acceptor. The resulting state $-(R_2Si)_x-(CH_2)_y-AT^-$ may be visualized as a logical "1" state. However, this state can have comparable energy to $-(R_2Si)_x^- -(CH_2)_y-AT$ state (logical "0"). Due to tunneling processes it is very likely that the $-(R_2Si)_x-(CH_2)_y-AT^-$ state will decay very rapidly into the $-(R_2Si)_x^- -(CH_2)_y-AT$ state and therefore lose its information.[87,92] Of course, one may argue that even modern semiconductor RAMs have to be refreshed every few milliseconds and a corresponding "refresh cycle" also could be provided in this case. One should be aware, however, that power consumption is also a factor of importance. For instance, while molecularly based computation might reasonably promise computation energies of ~50 kT/b handled. The acceptor ended polysilane approach makes little progress in energy per bit handled. The fundamental origin of this ineffectiveness is the parallel processing. For example, the need for 5000 electrons necessitates 5000 strands everywhere.

It is also important to select properly the excitation energy of the end acceptor substituents wavelength. A system designed so that $D \rightarrow A$ transfer of the switching substituents is driven by a wavelength λ_1, while $AT \rightarrow AT^*$ transfer is driven by λ_2, would have the desired property. The advantage of this scheme is that the light pulses are answers to the clock, power supply, and switching problems. With light, the fabrication problems of real devices appear to lie within understood polysilane chemistry, VLSI technology, and physics of polysilane polymers.

The general question of light-driven polysilane-based devices can be investigated with far shorter oligomers tethered at only one end, and without the need for microfabrication.

B. NONLINEAR POLYSILANE MATERIALS FOR INTEGRATED OPTICS

New polymers with unusual optical and electrical properties are predicted to have a high market potential in the future.[93] The combination of optical and electrical properties is especially attractive in modern electronics with long-term ambitions concerning all-optical devices.[94] Thus, material research using functionalized polymers with specially designed tailor-made properties becomes very important.[75] There are indications that for certain applications, polysilane materials will show significantly better performance than organic ones.[95] The reasons for interest in polysilane materials and some problems to be solved before they can be used commercially have been discussed recently in the Miller review.[96]

IV. CONCLUDING REMARKS

In this discussion, we have omitted reference to the challenging and interesting problems of polysilane chemistry and physics (for example, many aspects of photochemistry, photophysics, redistribution equilibria or conformational phenomena, and hole burning in the optical spectra of delocalized backbone states). Interested readers can find these topics in the Reference section. In spite of this, we hope that the present study demonstrates several potential applications for these interesting materials.

Molecular electronics is a challenge to all scientists. New developments will take different paths, but it is generally accepted that the main problems will not be solved in the near future. We should not forget that current and prospective silicon chip technology represents one of the most important technologies characterizing our modern way of life; therefore, my suggestion of polysilane-based MED is not an alternative system to compete with silicon electronics, but a supplement to it for future improvements.

ACKNOWLEDGMENTS

Particular thanks are given to Dr. Werner Marx (Max-Planck-Institut für Festkörperforschung, Stuttgart, Germany) for informative CA ONLINE services. I am also grateful to Dr. Slawomir Błoński, Dr. Boguslaw Dręczewski, Dr. Stanislaw Konieczny, and Professor Wieslaw Wojnowski for their contribution to the research reviewed here. Informal discussions during my stay at Max-Planck-Institut Für Festkörperforschung, with Dr. Erich Schönherr, Dr. Karlheinz Vogel, and Dr. Maria Wojnowski were helpful.

REFERENCES

1. **Smith, M. Z.,** *FORTRAN 77 A Problem-Solving Approach,* Houghton-Mifflin, Boston, 1985.
2. **de Solla Price, D. J.,** *Science Since Babylon,* Yale University Press, New Haven, 1961.
3. **de Solla Price, D. J.,** *Little Science — Big Science,* Columbia University Press, New York, 1963.
4. **Munn, R. W.,** Molecular electronics, *Chem. Br.,* 20(6), 518, 1984.
5. **Jech, Č.,** Molecular electronics, *Ustav Fyzi. Chem. Electrochem.,* CSAV, Prague, 1986.
6. **Roth, S.,** Poliene chains in molecular electronics, *Z. Phys. B,* 64, 25, 1986.
7. **Roth, S.,** Poliene chains in molecular electronics, *Synth. Met.,* 21, 51, 1987.
8. **Chiabrera, A., Di Zitti, E., Costa, F., and Bisio, G. M.,** Physical limits of integration and information processing in molecular systems, *J. Phys. D.,* 22, 1571, 1989.
9. **Aviram, A. and Ratner, A.,** Molecular rectifiers, *Chem. Phys. Lett.,* 29, 277, 1974.
10. **Miller, J. S.,** Molecular electronics?, *Adv. Mater.,* 2(8), 378, 1990.
11. **Miller, J. S.,** Molecular electronics?, *Adv. Mater.,* 2(10), 495, 1990.
12. **Miller, J. S.,** Molecular electronics?, *Adv. Mater.,* 2(12), 601, 1990.
13. **Roth, S.,** Switching processes in molecules and conducting polymers, *Phys. Scripta,* T29, 206, 1985.
14. **Slanina, Z.,** Thermodynamics of molecular electronics agents: the significance of highly mobile isomeric equilibria, *Thermochim. Acta,* 159, 77, 1990.

15. **Miller, R. D. and Michl, J.,** Polysilane high polymers, *Chem. Rev.*, 1359, 89, 1989.
16. **Lehn, J.-M.,** Perspectives in supramolecular chemistry — from molecular recognition towards molecular information processing and self organization, *Angew. Chem. Int. Ed. Engl.*, 29, 1304, 1990.
17. **West, R.,** Polysilanes, in *The Chemistry of Organic Silicon Compounds,* Patai, S. and Rappoport, Z., Eds., John Wiley & Sons, New York, 1989.
18. **Pitt, C. G.,** The conjugative properties of polysilanes and catenated σ-systems, in *Homoatomic Rings, Chains and Macromolecules of Main-Group Elements,* Rheingold, A. L., Ed., Elsevier, Amsterdam, 1977.
19. **Sandorfy, C. and Daudel, R.,** Sur la méthode des orbitales moléculaires appliquée a l'étude des liaisons σ, *C. R.*, 238, 93, 1954.
20. **McWeeny, R.,** The density matrix in many-electron quantum mechanics, *Proc. R. Soc.*, 253, 242, 1959.
21. **McWeeny, R.,** Some recent advances in density matrix theory, *Rev. Mod. Phys.*, 32, 335, 1960.
22. **Sinanoğlu, O.,** A principle of linear covariance for quantum mechanics and the electronic structure theory of molecules and other atom clusters, *Theor. Chim. Acta,* 65, 233, 1984.
23. **Sinanoğlu, O.,** On the algebraic construction of chemistry from quantum mechanics. A fundamental valency vector field defined on the euclidean 3-space and its relation to the Hilbert space, *Theor. Chim. Acta,* 65, 243, 1984.
24. **Sinanoğlu, O.,** Non-unitary classification of molecular electronic structures and other atom clusters, *Theor. Chim. Acta,* 65, 249, 1984.
25. **Sinanoğlu, O.,** Structural covariance of graphs, *Theor. Chim. Acta,* 65, 255, 1984.
26. **Sinanoğlu, O.,** Deformational covariance of graphs, *Theor. Chim. Acta,* 65, 267, 1984.
27. **Sinanoğlu, O.,** Hamiltonian as a Hessian on the Hilbert space, eigenvectors as critical points, and their relation to topological invariants in the variation method, *Theor. Chim. Acta,* 65, 271, 1984.
28. **Sinanoğlu, O.,** A theorem for qualitative deductions in organic or inorganic chemistry regarding the relative stabilities, distortions and reactions of molecules, *Chem. Phys. Lett.*, 103, 315, 1984.
29. **Sinanoğlu, O.,** New method for qualitative quantum chemical deductions on organic or inorganic molecules or clusters directly from structural formulas or ORTEP diagrams, *Theor. Chim. Acta,* 68, 251, 1985.
30. **Herman, A., Dręczewski, B., and Wojnowski, W.,** σ-Bond conjugation in polysilanes, a PES scaled free-electron approach for the interpretation of skeletal cleavage reactions, *J. Organomet. Chem.*, 251, 7, 1983.
31. **Herman, A., Dręczewski, B., and Wojnowski, W.,** The degree of σ-bond delocalization in linear and branched catenates of carbon and silicon, in *Proc. 6th Polish Organosilicon Symp.*, Dymaczewo (Poland), June 1983.
32. **Herman, A., Dręczewski, B., and Wojnowski, W.,** The degree of σ-bond delocalization in polysilanes and its influence on reactivity by through-bond interactions. The PES scaled Sandorfy C approach, *Chem. Phys.*, 98, 475, 1985.
33. **Herman, A., Dręczewski, B., and Wojnowski, W.,** ^{29}Si–N.M.R. chemical shifts of trialkoxysilylthio derivatives of permethylpolysilanes, *Z. Anorg. Allg. Chem.*, 551, 196, 1987.
34. **Herman, A., Dręczewski, B., and Wojnowski, W.,** σ-Bond conjugation in silicon-sulphur catenates, *J. Organomet. Chem.*, 339, 41, 1988.
35. **Herman, A.,** The Sandorfy "C" SCF CI(1) method for the interpretation of polysilane UV and PES spectra, in *Proc. 7th Polish Organosilicon Symp.*, Jachranka, Poland, 1986.
36. **Herman, A.,** An improved Sandorfy C method for permethylpolysilanes, *Chem. Phys.*, 122, 53, 1988.
37. **Bloński, S., Herman, A., and Konieczny, S.,** Photochemistry of permethylpolysilanes. Bond order changes accompanying excitation — SCF CI(1) Sandorfy C approach, *Spectrochim. Acta,* 45A, 747, 1989.

38. **Bloński, S., Herman, A., and Konieczny, S.,** Photochemistry of phenyl-substituted methylpolysilanes. Extended Pariser-Parr-Pople-Sandorfy C approach, *Spectrochim. Acta,* 45A, 753, 1989.
39. **Herman, A. and Bloński, S.,** Extended Pariser-Parr-Pople-Sandorfy C Method, *Spectrochim. Acta,* 44A, Software Survey Section II, 1988.
40. **Christoffersen, R. E.,** *Ab initio* calculations on large molecules, *Adv. Quantum Chem.,* 6, 333, 1972.
41. **Redondo, A., Goddard, W. A., III, McGill, T. C., and Surratt, G. T.,** Relaxation of (111) silicon surface atoms from studies of Si_4H_9 clusters, *Solid State Commun.,* 20, 733, 1976.
42. **Redondo, A., Goddard, W. A., III, McGill, T. C., and Surratt, G. T.,** Relaxation of (111) silicon surface atoms from studies of Si_4H_9 clusters, *Solid State Commun.,* 21, 991, 1977.
43. **Moskowitz, J. W., Topiol, S., Synder, L. C., and Ratner, M. A.,** Scaled single-zeta basis set for use with a silicon effective potential: generalized valence bond description of disilane, *Int. J. Quantum Chem.,* 19, 131, 1981.
44. **Palke, W. C.,** Double bonds are bent equivalent hybrid (banana) bonds, *J. Am. Chem. Soc.,* 108, 6543, 1986.
45. **Messmer, R. P.,** Polyhedral hybrid approach to bonding in molecules containing transition metals, *Chem. Phys. Lett.,* 132, 161, 1986.
46. **Garrison, B. J. and Goddard, W. A., III,** Dissociation energetics of fluorosilylidyne (SiF) systems of relevance to etching reactions, *J. Chem. Phys.,* 87, 1307, 1987.
47. **Teramae, H.,** *Ab initio* studies on the silicon compound: on the electronic structure of disilene reconsidered, *J. Am. Chem. Soc.,* 109, 4140, 1987.
48. **Pople, J. A. and Santry, D. P.,** A molecular orbital theory of hydrocarbons. I. Bond delocalization in parafins, *Mol. Phys.,* 7, 269, 1963–1964.
49. **Pople, J. A. and Santry, D. P.,** A molecular orbital theory of hydrocarbons. II. Ethane, ethylene and acetylene, *Mol. Phys.,* 9, 301, 1965.
50. **Dewar, M. J. S.,** σ-Conjugation and σ-aromaticity, *Bull. Soc. Chim. Belg.,* 88, 957, 1979.
51. **Dewar, M. J. S.,** Aspects of cyclic conjugation, *Pure Appl. Chem.,* 52, 1431, 1980.
52. **Dewar, M. J. S.,** Chemical implications of σ-conjugation, *J. Am. Chem. Soc.,* 106, 669, 1984.
53. **Spangler, D. and Christoffersen, R. E.,** Development of basis sets for molecular calculations. Comparison of tetrahedral Gaussian lobe functions and Cartesian Gaussian basis sets, *Int. J. Quantum Chem: Quantum Biol. Symp.,* 5, 127, 1978.
54. **Toman, J. J., Frost, A. A., Topiol, S., Jacobson, S., and Ratner, M. A.,** *Ab initio* studies using the FSGO-pseudopotential method, of the electronic structure in group IV cluster molecules $M(LH_3)_4$; M,L=C,Si,Ge, *Theoret. Chim. Acta,* 58, 285, 1981.
55. **Stüger, H., Hengge, E., and Janoschek, R.,** PE and UV spectra of linear and cyclic polysilanes and assignment by *ab initio* calculations, *Phosphorus, Sulfur Silicon,* 48, 189, 1990.
56. **Gaspar, P. P.,** Silylenes, in *Reactive Intermediates — A Serial Publication,* Vol. 3, Jones, M., Jr. and Moss, R. A., Eds., John Wiley & Sons, New York, 1985.
57. **Kipping, S.,** Complex silicohydrocarbons $[SiPh_2]_n$, *J. Chem. Soc.,* 125, 2291, 1924.
58. **Burdett, J. K.,** From bonds to bands and molecules to solids, *Prog. Solid State Chem.,* 15, 173, 1984.
59. **Albright, T. A., Burdett, J. K., and Whangbo, M.-H.,** *Orbital Interactions in Chemistry,* John Wiley & Sons, New York, 1985.
60. **Herman, A.,** Properties of conducting σ-electron systems deduced from UV photoelectron spectroscopy of model compounds, *Summer School of Conducting Polymers* (poster session), Max-Planck-Institut für Festkörperforshung and Polish Academy of Sciences, Sulejów, Poland, 1988.

61. **Herman, A.,** Properties of conducting σ-electron systems deduced from UV photoelectron spectroscopy of model compounds, in *8th Polish Organosilicon Symp.,* Sulejów, Poland, 1989.
62. **Brédas, J. L., Thémans, B., André, J. M., Chance, R. R., and Silbey, R.,** The role of mobile organic radicals and ions (solitons, polarons and bipolarons) in conjugated polymers, *Synth. Met.,* 9, 265, 1984.
63. **Hirschfelder, J. O.,** Where are laser molecule interactions headed?, in *Lasers, Molecules and Methods,* Hirschfelder, J. O., Wyatt, R. E., and Coalson, R. D., Eds., John Wiley & Sons, New York, 1989.
64. **Lauchlan, L., Etemad, S., Chung, T. C., Heeger, A. J., and MacDiarmid, A. G.,** Photoexcitations in polyacetylene, *Phys. Rev. B,* 24, 3701, 1981.
65. **Fesser, K., Bishop, A. R., and Campbell, D. K.,** Optical absorption from polarons in a model of polyacetylene, *Phys. Rev. B,* 27, 4804, 1983.
66. **Alred, A. L. and Van Beek, D. A., Jr.,** Electrical conductivity of some permethylpolysilanes doped with arsenic pentafluoride, *Polyhedron,* 10, 1227, 1991.
67. **Patil, A. O., Heeger, A. J. and Wundl, F.,** Optical properties of conducting polymers, *Chem. Rev.,* 88, 183, 1988.
68. **Balaji, V. and Michl, J.,** Singlet excitation in polysilanes: *ab initio* calculations on oligosilane models, *Polyhedron,* 10, 1265, 1991.
69. **Kepler, R. G., Zeigler, J. M., Harrah, L. A., and Kurtz, S. R.,** Photocarrier generation and transport in σ-bonded polysilanes, *Phys. Rev. B,* 35, 2818, 1987.
70. **Henge, E.,** Properties and formation reactions of cyclic polysilanes, in *Homoatomic Rings, Chains and Macromolecules of Main-Group Elements,* Rheingold, A. L., Ed., Elsevier, Amsterdam, 1977.
71. **Brédas, J. L., Thémans, B., André, J. M., and Piela, L.,** Theoretical studies of localization phenomena in excited states of polyene chains, *Bull. Soc. Chim. Belg.,* 92, 1, 1983.
72. **Soos, Z. G., Hayden, G. W., and McWilliams, P. C. M.,** Pariser-Parr-Pople approach to σ-conjugated polysilane chains, *Polym. Prepr.,* 31, 286, 1990.
73. **Kepler, R. G. and Soos, Z. G.,** Electronic excitations of poly(methylphenylsilane) films, *Phys. Rev. B,* 15, 12530, 1991.
74. **Andre, J. M., Morley, J. O., and Zyss, J.,** From quantum chemistry to organic optical signal processing: a computer-aided molecular engineering approach, in *Molecules in Physics, Chemistry, and Biology,* Vol. 2, Maruani, J., Ed., Kluwer Academic Publ., Dodrecht, 1988.
75. **Andre, J. M. and Delhalle, J.,** Quantum chemistry and molecular engineering of oligomeric and polymeric materials for optoelectronics, *Chem. Rev.,* 91, 843, 1991.
76. **Chemla, D. S. and Zyss, J., Eds.,** Nonlinear optical properties of organic molecules and crystals, in *Quantum Electronics — Principles and Applications,* Academic Press, Orlando, 1987.
77. **Raabe, G. and Michl, J.,** Multiple bonding to silicon, *Chem. Rev.,* 85, 419, 1985.
78. **Schoeller, W. W. and Dabisch, T.,** Bonding in cyclosilanes. Effect of orbital nonhybridization, *Inorg. Chem.,* 26, 1081, 1987.
79. **Allred, A. L., Ernst, C. A., and Ratner, M. A.,** Conformations and long-range interactions in permethylpolysilanes, in *Homoatomic Rings, Chains and Macromolecules of Main-Group Elements,* Rheingold, A. L., Ed., Elsevier, Amsterdam, 1977.
80. **Michl, J., Downing, J. W., Karatsu, T., McKinley, A. J., Poggi, G., Wallraff, G. M., Sooriyakumaran, R., and Miller, R. D.,** Solution photochemistry of poly(dialkylsilanes): a new class of photoresists, *Pure Appl. Chem.,* 60, 959, 1988.
81. **Zeigler, J. M.,** Polysilanes — prototypical σ-conjugated polymers, *Synth. Met.,* 28, C581, 1989.

82. **Miller, R. D.,** Radiation sensitivity of soluble polysilane derivatives, in *Silicon — Based Polymer Science: A Comprehensive Resource,* Zeigler, J. M. and Fearon, F. W. G., Eds., Advances in Chemistry Series, No. 224, American Chemical Society, Washington, D.C., 1990.
83. **Ramsey, B. G.,** Electronic transitions of polysilanes and their photochemistry, *J. Organomet. Chem.,* 67, C67, 1974.
84. **Carter, F. L.,** Molecular electronic devices, today's dream — tomorrow's reality, *BIOTECH '84 USA,* On-line Publ., Pinner, U.K., 1984.
85. **Bertan, D. N.,** Controlled electron transfer for molecular electronics, *Mol. Cryst. Liq. Cryst.,* 190, 85, 1990.
86. **Hopfield, J. J., Onuchic, J. N., and Bertan, D. N.,** Electronic shift register memory based on molecular electron-transfer reactions, *J. Phys. Chem.,* 93, 6350, 1989.
87. **Newton, M. D.,** Quantum chemical probes of electron — transfer kinetics: the nature of donor-acceptor interactions, *Chem. Rev.,* 91, 767, 1991.
88. **Harrod, J. F., Mu, Y., and Samuel, E.,** Catalytic dehydrocoupling: a general method for the formation of element-element bonds, *Polyhedron,* 10, 1239, 1991.
89. **Eaborn, C. and Batt, R. W.,** Synthesis and reactions of the silicon-carbon bond, in *Organometallic Compounds of the Group IV Elements. The Bond to Carbon,* MacDiarmid, A., Ed., Marcel Dekker, New York, 1968.
90. **Speier, J. L.,** Homogeneous catalysis of hydrosilylation by transition metals, *Advan. Organometal. Chem.,* 17, 407, 1979.
91. **Lukevits, E. Y. and Voronkov, M. G.,** *Organic Insertion Reactions of Group IV Elements,* Plenum Press, New York, 1966.
92. **Mehring, M.,** Concepts of molecular information storage, in *Electronic Properties of Conjugated Polymers III,* Kuzmany, H., Mehring, M., and Roth, S., Eds., Springer Series in Solid-State Sciences, Vol. 91, Springer-Verlag, Berlin, 1989.
93. **Groh, W., Lupo, D., and Sixl, H.,** Polymer optical fibers and nonlinear optical device principles, *Angew. Chem. Int. Ed. Engl.,* 28, 1548, 1989.
94. **Zyss, J.,** Nonlinear organic materials for integrated optics: a review, *J. Mol. Electron.,* 1, 25, 1985.
95. **Mignani, G., Krämer, A., Puccetti, G., Ledoux, I., Soula, G., Zyss, J., and Meyrueix, R.,** A new class of silicon compounds with interesting nonlinear optical effect, *Organometallics,* 9, 2640, 1990.
96. **Miller, R. D.,** Polysilanes — A new look at some old materials, *Angew. Chem. Int. Ed. Engl.,* 28, 1733, 1989.

Chapter 3

INFORMATION PROCESSING AND COMPUTING DEVICES BASED ON BIOMOLECULAR NONLINEAR DYNAMIC SYSTEMS

N. G. Rambidi, D. S. Chernavskii, and V. I. Krinsky

TABLE OF CONTENTS

0-8493-8061-8/93/$0.00 + $.50

I. INTRODUCTION: MOLECULAR AND BIOMOLECULAR COMPUTING: CONTEMPORARY STATE OF THE ART

At the annual meeting of the American Physical Society in December 1959, Feynman, a famous physicist, in his remarkable talk, "There Is Plenty of Room at the Bottom",[1] proposed the use of molecular level phenomena and microdevices designed and constructed of atomic (molecular) primitives. He suggested a wide range of applications for these devices and essentially novel results that might be obtained following this direction.

He prophetically declared that the capabilities of these devices would be highly successful in processing and manipulating the vast data arrays, solving a number of complex research problems, achieving more profound understanding of the information of biological processes, etc.

In the late 1950s, electronics had just started on the way to using a new planar technology for building integrated circuits. Therefore, the prescient ideas of Feynman had a profound influence on the development of a number of fields of science and technology (see, for instance Krumhansl and Pao[2] and Drexler[3]), primarily informatics and computing science.

Nowadays, engineering and technology, transport and communication, planning, accounting, and other structures providing the existence and development of human society are inconceivable without utilization of the "information industry" based on the means and methods of the present-day electronics. The great progress made in this field was due to employment of semiconducting inorganic materials and planar technology in manufacturing submicron-level devices. This technological approach has made personal computers with a RAM capacity of several megabytes widely available and has led to production of computing complexes with an information processing speed of more than 100 million floating-point instructions per second (MIPS). Though the capabilities of planar semiconductor technology are far from being exhausted, the physical limits are seen now; and today many groups of scientists are searching for novel solutions and new materials that would widen the potentialities of electronics.

One of the trends in information processing based on processes of molecular transformations in organic and organoelemental compounds is to use the name of "molecular electronics". In general, the spectrum of new physicochemical phenomena emerging when passing from conventional semiconductor logical primitives to ordered molecular assemblies (organic crystals, liquid-crystalline phases, Langmuir-Blodgett films, etc.) appears to be rather rich. Condensed organic and organoelemental substances are characterized by weak intermolecular interaction that allows many individual molecular properties to be retained. At the same time, the inherent properties of supramolecular structures and independent large polymeric molecules represent a curious interplay of the properties of individual molecules (molecular fragments) and collective properties of condensed substances.

Moreover, complex nonlinear dynamic processes at the molecular level occurring in chemical and biochemical systems would result, under certain conditions, in self-organization of the system. In this case one would observe the appearance of stationary spatially nonuniform or temporary oscillating microstates (e.g., heterogeneity of component concentration) in the system. It is noteworthy that these processes form the basis for information processing by simple biological systems. All these characteristics allow one to use organic and organoelemental materials to develop information processing devices having capabilities essentially different from those of conventional computing systems.

The main research trends in the field of molecular electronics are presented in Table 1. More detailed information on most of them can be found in the references in this table.[4-12] Of all the trends, we shall consider only the problem of designing and constructing information processing devices at the molecular level (i.e., the use of individual molecules or their assemblies as primitives for information processing devices), such as molecular and biomolecular computing. This work is an attempt to present an overview of the relevant

TABLE 1
Major Trends in the Field of Molecular Electronics

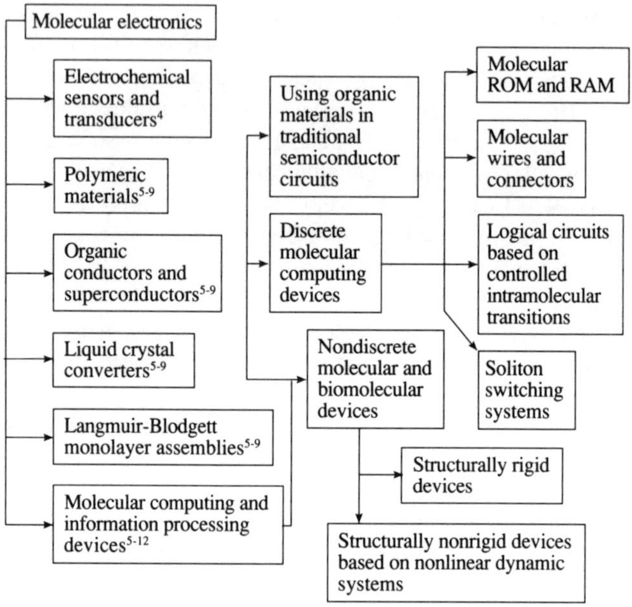

works published in the last decade. The major trends that can be distinguished in these works are

1. Studies of the sets of logical states which characterize a molecular or supramolecular structure as a discrete or continuous automaton
2. Molecular design, i.e., theoretical elaboration of the structures of molecular systems which possess the properties of logical elements (molecular primitives)
3. Experimental study of information processing at the molecular level, and assessment of correspondence between molecular characteristics and properties found and the predicted ones
4. Development of the architecture of information processing systems optimally utilizing the peculiarities of molecular primitives
5. Analysis of technological potentialities of manufacturing molecular information processing devices

At present the development of molecular tools for information processing proceeds along two lines:

1. Designing digital (discrete) molecular devices in which traditional semiconducting elements are substituted by their molecular analogs

(molecular diode, molecular transistor, etc.) while the architecture of the device per se remains conventional
2. Building up nondiscrete (pseudoanalog) molecular devices employing the principles of information processing at the molecular level in simple biological systems having flexible adaptable structures, i.e., functionally flexible devices

In evaluating the prospects for implementation of molecular primitives it would be desirable to estimate their anticipated advantages.

One of the main features of discrete molecular computing and information processing devices according to Airapetyanz[11] is that they function at the bridge between macro- and microproperties. Molecular devices are large enough to have an inner structure of sufficient complexity. At the same time they are small enough to operate by single microparticles (electrons, ions, etc.). In these devices the number of particles involved in the interaction process is strictly defined, and therefore they are not subject to shot noise inherent in macroscopic electron devices. This, in fact, allows the implementation of a discrete multiplication of signal carriers that is equivalent to noise-free amplification at room temperature.

Airapetyanz[11] also concluded the possibility of constructing molecular computing devices having elements possessing minimum switch energy.

Thus it should be noted that discrete molecular primitives allow for improvement of a number of information processing device characteristics as compared with similar characteristics of silicon-based devices. At the same time, molecular primitives should have some qualitatively different properties other than those of the silicon-based ones.

Characteristics of molecular circuits that may be considered to be improved due to the use of molecular primitives are

1. Higher degree of integration, although the estimated difference in integration between molecular and semiconductor based devices is not large
2. Considerably lesser switching energies
3. Enhanced stability of circuits with respect to radiation, especially while increasing the degree of integration

Among qualitatively new characteristics of molecular primitives the following should be named:

1. Complete identity of characteristics of the molecular primitives, unlike the spread of characteristics of semiconductor-based ones due to their technological failures
2. Shot noise free, one-electron processes that are typical of molecular objects

3. Specific molecular mechanisms of signal transfer allowing elaboration of more logically complicated primitives than the semiconductor-based ones

In current literature it is widely believed that employment of molecular primitives would provide an essential increase in the integration level up to $10^{15}/cm^3$ (for details see Rambidi[10]) and a multifold increase of the processing speed. Indeed, there is no doubt that the small (up to several tens of nanometers) size of molecules predetermines a high density of elements. However, a comparative analysis of characteristics of molecular systems and semiconductor integrated circuits implies that the advantages of the former are not large.

Even according to most optimistic prognoses, the first molecular information processing devices are not liable to appear until the late 1990s. Therefore, when estimating the advantages of molecular primitives, it seems reasonable not to use current semiconductor devices for comparison but instead those that are to appear in the next decade.

Probably a typical representative of the present-day integrated very large-scale integration (VLSI) chip may be the Motorola 68020 monocrystal processor with a processing speed of 2.5 million instructions per second.[13] It contains 192,000 elements located on an 80-mm^2 crystal. It is believed[14] that in the next 10 years the increasing process of microminiaturization of semiconductor devices will result in reduction of the size of elements to the physical limit of about 0.25 μm. At the same time, according to prognoses[14,15] the employment of novel technologies (such as ion resistorless lithography) will open up the possibilities of constructing semiconductor devices having characteristic dimensions up to 0.01 to 0.025 μm. Studies along this line have already been undertaken. For example, Texas Instruments Company (U.S.) has initiated a program which envisages development of novel promising primitives to be used in quantum devices of nano-scale size and high speed processing in the middle of the next decade.[16]

It might be suggested[10] that in order to avoid undesirable quantum effects that reduce the reliability of molecular logical elements, one should use polyatomic molecules as primitives.

Large biological molecules such as proteins are built of several (from one to four) structural domains containing 100 to 250 amino acid groups and forming a globule with a diameter of about 25 Å.[17] Hence, the average size of the molecular logical element may be taken to be of the order of 100 Å, which is close to the size of future promising semiconductor elements.

As for memory devices, the physical limit of the memory-storage elements on magnetic domains is of the order of 50 nm.[14] This allows construction of memory storage devices with a packing density of 4×10^{10} elements per square centimeter. We shall show below that element density of molecular memory devices may reach 10^{10} to 10^{11} elements per square centimeter.

The processing speed of present-day semiconductor devices is characterized by a value of 5×10^{-10} s, the physical limit being 10^{-14} s.[14] In the case of molecular elements the speed of processing can be limited by the time of light quantum absorption by the molecule in ground electron state taken as 10^{-15} s.

Thus, it may be concluded that using molecular primitives ought to improve to some extent the functional characteristics of the devices. The estimates also suggest that due to further progress in semiconductor VLSI devices, the differences in size of molecular and semiconductor primitives should not be great. Nevertheless, in some special cases implementation of molecular primitives will probably make it possible to substantially reduce the size of the constructed devices or to increase their complexity under reasonable dimensions. It should also be kept in mind that the comparison is based on formal analysis of the anticipated characteristics and does not take into account such factors as technological gaps, problems of raw material production, material impurity, etc.

It should be noted that discrete computing devices constructed on the molecular level and using the planar technology could suffer from the nightmare of interconnections while increasing the degree of integration.

It should be emphasized that the development of discrete molecular computing systems is based on the use of conventional architecture inherent in present-day semiconductor computers. This implies that the classes of problems to be solved by discrete molecular computers should fully coincide with those solved currently by the available semiconductor digital devices. At the same time there exist other classes of problems that cannot effectively be solved by the presently functioning semiconductor digital computers based on primordial von Neumann background and rigidity of structural organization.

Molecular electronics started as an attempt to use new physical entities (new materials) to obtain advanced operational characteristics of numerical computing devices. However, later it appeared that there was another objective as important as the initial one. It was solving problems of high computational complexity (see, for instance, Andrew[18]).

These problems embrace:

1. Recognition and analysis of patterns, screens, and situations (analysis of ecological, economic, and meterological satellite information, microscopic patterns for fast diagnosis in medicine, material science, etc.)
2. Search for the best solution of problems described by complicated tree diagrams (e.g., a choice of the shortest route of a robot agent)
3. Search for behavioral solutions including decision making under uncertainties of environmental factors (control of autonomous surface travel facilities and similar problems)

4. Study of the temporal evolution of complex multifactorial systems (solution of economic, sociological, and management problems)
5. Solving nonlinear differential equations with arbitrary boundary and initial conditions

The theory of computation operates on the notion of computational complexity that is described by the so-called signal functions.[19] Temporal and capacity signal functions are defined as:

$$T_M(n) = \max t_M(p) \quad \text{for all words p of length n of Turing machine M}$$

$$S_M(n) = \max s_M(p) \quad \text{for all words p of length n of Turing machine M}$$

Here $t_M(p)$ and $s_M(p)$ are the number of steps made by machine M during computational process and memory size used in the computations, respectively.

From the point of view of computational complexity, problems can be classified as those of polynominal and exponential complexity. A problem falls into the polynominal complexity class if the number of resources required to solve it increases as a low order polynominal function of problem size. A problem is in the exponential complexity class if the resources required to solve it increase at least exponentially with an increasing problem size, for example, 2^n.

In practice, only polynominal complexity problems are effectively solvable while exponential complexity problems entail combinatorial explosion.

In recent years there has been a growing interest in the practical possibility of building nondiscrete (pseudoanalog) molecular devices capable of solving problems of high complexity (in more details see References 20–26). They are expected to be based on general principles of information processing at the molecular level realized in simple biological systems, i.e.:

1. Giant parallelism of information processing
2. Processing mechanisms based on complicated nonlinear dynamics
3. High efficiency of information transformations
4. Considerable informational complexity of computational (pseudoelementary) primitives
5. Effective mechanisms of information processing
6. Ability for variation and evolution of the molecular components of information processing devices including the ability for evolutionary learning

The functioning processes of complex organic molecules and simple biological structures correspond to logical functions equivalent to a large set of simple logical operations (such as AND, NAND, NOR, etc.) executed by semiconductor primitives.

For example, within the time of about 0.1 msec the enzyme molecule recognizes the substrate molecule, distinguishing it from other similar molecules in the enzyme environment, adjoins the substrate molecule promoting the substrate-product reaction, and then releases the molecules of the products.[20] Energy dissipation corresponding to one logical step of the enzymatic reaction is typically only 10 to 100 kT per step.

High complexity of logical functions corresponding to the processes realized by large organic molecules and simple biological structures allows one to expect that they may be utilized for fabrication of information-processing devices able to solve problems significantly more complicated than those solved by present-day computers.

Conrad[20] conducted a detailed study of computational capabilities arising from the use of large organic molecules and simple biological structures as primitives. In this investigation he was guided by his own axiomatic trade-off principle which states that a system cannot be effectively programmable, amenable to evolution by variation and selection, and computationally effective at the same time.

Without trying to prove this principle Conrad supported it by the following three statements:[21]

1. A universal programmable system must be able to solve any problem for any physical situation and therefore cannot be as effective as the system specially trimmed for a particular task.
2. A rigid program implies a sequential mode of operations, i.e., ineffective use of machine resources.
3. The program may be damaged even by a minor change; therefore, a well-programmable system does not allow a step-by-step variation of functions together with gradual alteration of the structure (program).

The design principles for current information processing semiconductor devices are the rigid structural organization combined with effective programmability. At the same time large organic molecules involved in the process of vital activity corresponding to complex sequences of logical operations represent special-purpose adaptable structurally flexible systems with very high performance.

Functionally flexible molecular systems can mimic the capabilities of biological systems with regard to their adaptability to environmental conditions. Therefore, they may appear to be more promising for solving problems of high computational complexity than structurally programmable computing machines.

These features of functionally adaptable molecular-level information-processing devices allowed Conrad[20] to make a rather important conclusion: "From the standpoint of constructing artificial computers out of new materials, the implication is clear: what is needed is not a von Neumann's computer fabricated from alternative materials, but new computer designs to address computing needs not efficiently met by today's von Neumann's machines.''

Recently Conrad discussed in more detail very important features of biomolecular computing — the profound meaning of the trade-off principle and some other important problems.[28-30]

In our publications[11,12,23-27] an attempt was made to develop an approach to biomolecular computing consistent with Conrad's ideas but in some aspects with more engineering.

It should be stressed that functionally flexible information processing devices offer a new perspective for the implementation of data processing principles at the molecular level. If molecular level computers with flexible intrinsic structures featuring adaptability and evolution principles could be designed and constructed, they essentially would differ from the perspective of future systems employing semiconductor primitives. They would be complementary rather than competitive to digital computers by their functions and capabilities.

Different possibilities of fabricating molecular information processing devices will be examined below. The limits of this chapter prevent us from considering all relevant works available in the literature. Therefore, our interest will be focused on:

1. Design principles which may underlie the molecular information processing devices, and general principles of molecular primitives elaboration
2. Basic directions in construction of digital devices employing molecular primitives
3. Possibilities of fabricating analog and functionally flexible information processing systems

II. DISCRETE MOLECULAR COMPUTING DEVICES

A. THE POSSIBILITIES OF IMPLEMENTING BASIC ELEMENTS FOR DISCRETE COMPUTING AND INFORMATION PROCESSING SYSTEMS AT THE MOLECULAR LEVEL

The possibilities of fabricating numerical (discrete) computing and information processing devices have been the subject of extensive discussions in literature in recent years.[5-12] Generally, the authors of these works assumed (explicitly or implicitly) that the base for the molecular computer would be the consecutive structurally rigid architecture similar to the von Neumann one. There were only a few works which discussed the possibility of parallel mode of information input and information processing, the usage of cellular automatons approach, etc.

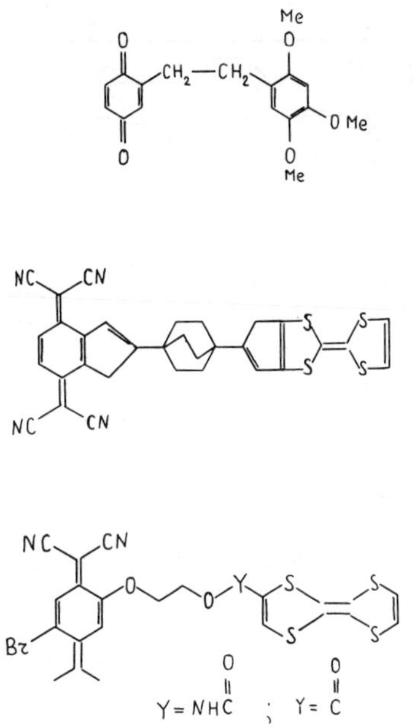

FIGURE 1. Molecules built out of electron-donor and electron-acceptor fragments — molecular rectifiers (see explanation in the text).

Major interest has been focused on the physical foundation and the ways of implementing basic elements of computing and information processing devices at the molecular level.

The first work dealing with primitives at the molecular level was "Molecular Rectifiers" by Aviram and Ratner.[31] The molecular system suggested by them contains two fragments: one being the electron acceptor and the other, the electron donor conjugated system confined between two electrodes. It is clear that for such a molecule to be stable, the donor and acceptor groups must be separated from each other. It can be done by inserting a hydrocarbon bridge between donor and acceptor fragments; that would result in a conjugation breakdown and in a possibility for the electron to move by tunneling, along the molecular system in the direction:

$$\text{cathode} \rightarrow \text{acceptor} \rightarrow \text{donor} \rightarrow \text{anode}$$

Examples of the molecules having rectifying properties are presented in Figure 1.

Among them is a simple molecule in which the *n*-benzoquinone group is an acceptor and methyl ether of oxyhydroquinone is a donor of the electron. Figure 1 illustrates an analogous molecule with the known acceptor and donor

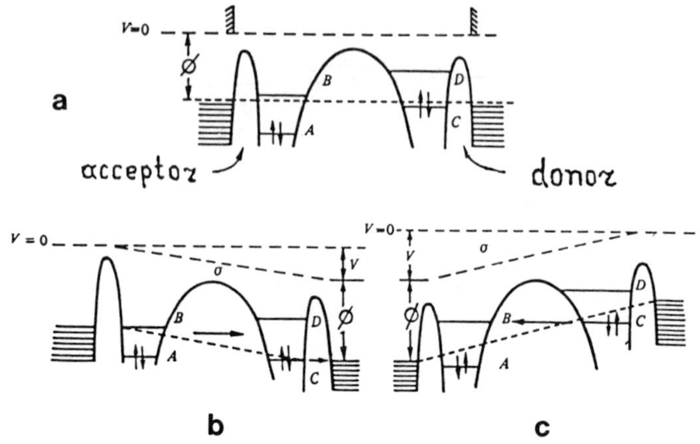

FIGURE 2. The simplified model of electron tunneling in molecular rectifier (see explanation in the text).

fragments, tetracyanoquinodimethane (TCNQ) and tetrathiafulvalene (TTF), respectively.[31] In this molecule the donor and acceptor are coupled not by one bridge but by a system of three methylene ($-CH_2-$) bridges.

Let us examine in detail the effect of asymmetric conductivity for the molecule shown in Figure 1. For the molecule to possess rectifying properties, it is necessary that the upper unoccupied or partially occupied electronic level B of the molecule acceptor fragment should lie somewhat higher than the Fermi level of the electrode as shown in Figure 2, and the upper occupied level of molecule C donor fragment should lie below the Fermi level of the electrode. When the applied electric field becomes strong enough to have the cathode level overlapped with the acceptor level (Figure 2), there arises a possibility for the electron to pass from the cathode to the acceptor. The threshold of this process depends on the energy level E_V and the work function of metal (φ). Similarly, for the donor fragment, electron transfer from the donor orbital C to anode would occur if the applied voltage $V > I - \varphi$ where I is the donor ionization potential. Electron transfer from acceptor to donor occurs under the action of the electrical field due to nonelastic tunneling. The charged acceptor contains an electron on the upper orbital B (the basic vibrational state) that without any energy loss passes by tunneling into the excited vibrational donor state to the upper orbital C, since level B lies higher than level C; and if further passes to the basic level C. The transition from level B to level C is irreversible since B lies higher than C.

If the field polarity is changed to the opposite sign, the situation is different (Figure 2). Tunneling will be possible when the Fermi level of the metal drops below level A. The additional conductivity mechanism can come into operation provided there is tunneling from level C of the donor to level B of the acceptor. Accordingly, the probability of reverse electron transition will decrease in comparison with direct transition.

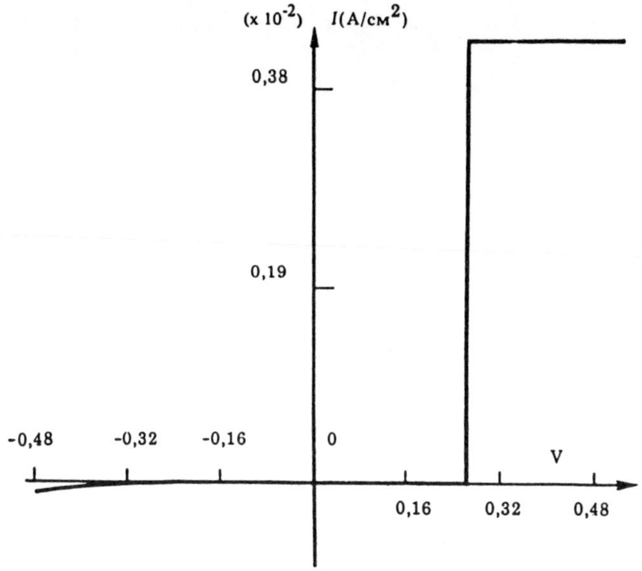

FIGURE 3. The voltage-current characteristic for a molecular rectifier.

The voltage-current characteristic describing the process of intramolecular electron transfer can be calculated using analytical description of transition processes. Not only the mechanisms discussed above might be considered, but also other possible mechanisms as well. One of them is direct tunneling from one electrode to another through the σ bond as a barrier. Probably the contribution of such direct transition will be small as shown by Kuhn et al.[32,33] who measured the conductivity of monomolecular Langmuir-Blodgett film assemblies built of fatty acid molecules. Figure 3 shows a voltage-current characteristic calculated in Aviram and Ratner[31] which really describes the rectifying intramolecular effect. Of course, when speaking about electric current, one should mean a rather large ensemble of similar molecules.

The ideas of Aviram and Ratner[31] on the fabrication of molecular systems with asymmetric electron conductivity had prompted researchers to synthesize such molecules. Panetta et al.[34] and Metzger and Panetta[35,36] were able to synthesize candidates for these molecules (see Figure 1). They also synthesized molecules described in Aviram and Ratner[31] but possessing amphiphilic properties, and obtained ordered mono- and multimolecular LB structures.[35,36] However, they failed to measure until now the electronic conductivity of the molecules suggested in Aviram and Ratner.[31]

Later Aviram et al.[37] suggested building a molecular switch on the basis of semiquinone-type molecules. These molecules (Figure 4) may occur in two stable states that differ in arrangement of hydrogen atoms. The stable states are separated by a comparatively low energetic barrier that makes possible

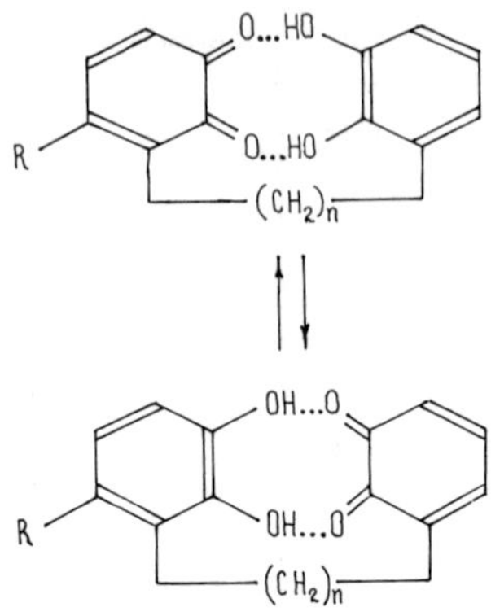

FIGURE 4. A molecule of semiquinone type as an example of switching molecules.

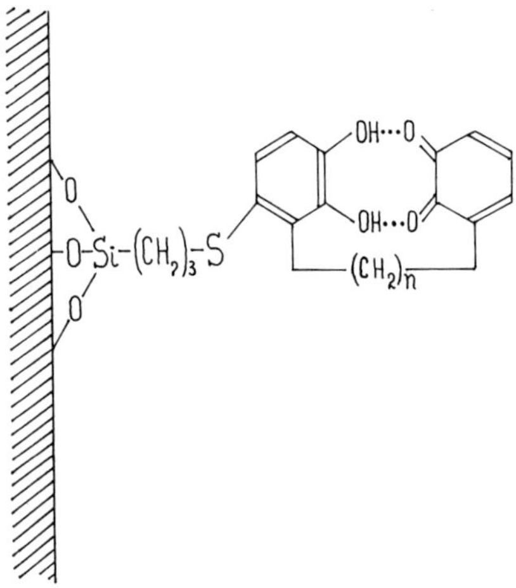

FIGURE 5. The simplified model of the molecular switch.

field-driven proton tunneling. Thus, the semiquinone-type molecule is a simple molecular switch controlled by the electric field.

Proceeding from their concept Aviram et al.[37,38] performed a number of experiments aimed at developing an operational model of a molecular switch. By using chemical reactions on the aluminum surface, they were able to bind the semiquinone molecules with the surface through intermediate molecular bridges (Figure 5). The molecules in this case are arranged perpendicularly to the surface and have different absorption spectra for different positions of hydrogen nuclei with respect to the molecule skeleton. This allows integral estimation (for a large number of molecules on the surface) of preferential positions of hydrogen atoms in the structure of semiquinone molecules.

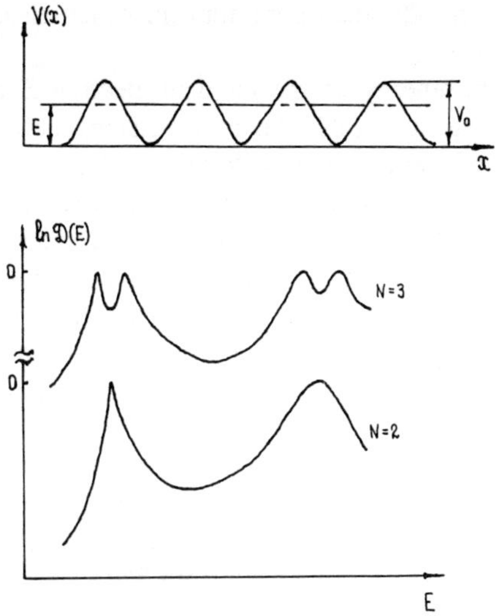

FIGURE 6. The electron tunneling through the sequence of one-dimensional barriers (see explanation in the text).

Later Aviram et al.[39] synthesized an analogous molecular system on a gold surface and demonstrated the possibility of controlling the state of an individual molecule by using a scanning tunneling microscope.

Several fundamental approaches to building discrete molecular primitives were proposed by Carter[40] and Carter et al.[41] In general, research has developed in two major directions:

1. Using the resonance electron tunneling mechanism in the system of sequentially arranged potential barriers, and control of the system by shift of the level position in one of the potential wells (concept of control groups)
2. Using soliton mechanisms of excitation transfer to change the electron structure of molecular systems for switching molecules from one stable state to another (concept of switching molecules)

The first type of problems is related to electron tunneling through the sequence of N quasi-one-dimensional barriers of arbitrary shape V(x) separated by (N − 1) potential wells, provided that at $X \to \pm\infty$ $V(x) \to 0$. The qualitative features of tunneling do not depend on the shape of the potential energy barriers. They were analytically calculated: (1) for barriers of δ function[42] type and (2) in the pseudoclassical approximation, for barriers of arbitrary shape in 1962 by Pshenichnov;[43] who showed that electron tunneling might be anomalously large under certain conditions.

A detailed analysis suggested[43] that at $0 < E < V_0$ (Figure 6) pseudostates may arise in each of the minima, whose spectra may roughly be represented as the spectrum of an individual minimum, in which all levels are split into (N − 1) components.

Moreover, if the energy of the incident electron E coincides with the energy of one of the sublevels, the transmission coefficient will show a nonmonotonous dependence on E, with (N − 1) maxima where the transmission coefficient will equal unity. Thus there arises a possibility for easy electron passage through a sequence of barriers. At the same time, variation in the shape of even one of the barriers changes the position of the levels (with respect to the energy of the incident particle), and the transmission coefficient will sharply decrease. Therefore, this system may provide both long-distance signal transmittance and effective interruption of the signal.

It should be noted that the theoretical calculations performed by Pshenichnov[43] so far have not been experimentally confirmed for any molecular system. In 1973, however, similar calculations were done by Tsu and Esaki[44] for multilayer semiconductor heterostructures, and the results of calculations supported Pshenichnov conclusions.[43] Tsu and Esaki were able to obtain the dependences of the transmission coefficient on the energy of the tunneling electron. They exhibited both a sharp increase in transmission, when electron energy coincided with one of the levels of the system of square-topped barriers, as well as the peak splitting. Later results of Tsu and Esaki[44] received a convincing experimental support, particularly, for $GaAs_x$–$Ga_{1-x}As$ heterostructures.[45]

In practice, the situation considered in Reference 43 is complicated by the presence of the external electrical field,[42] which results in changing the barrier shape V_x and decreasing the transmission coefficients for electron tunneling below unity. Therefore, for resonance tunneling to be realized the initial potential should be chosen so that the transmission coefficients were close to unity in the presence of the superimposed electric field.

The possibilities of controlling molecular systems by means of resonance tunneling can be demonstrated by the examples of NAND and NOR elements suggested by Carter et al.[40,41] It should be noted that either one of these elements will be enough to build any logical circuit. Figure 7 illustrates a hypothetical structure of a NAND element. The positively charged aromatic heterocyclic groups of this element represent potential wells for the electron, and the connecting diazogroups are potential barriers. In Figure 7 the depth of the potential well is determined by the charge of the nitrogen atom. Of four controlling groups, two (CG_1 and CG_3) keep the charge of nitrogen unchanged, while in the CG_2 and CG_4 groups structural rearrangement (because of the charge changing on the nitrogen atom) is possible under the action of some physical factor (for instance, light absorption). Since neutralization of the aromatic group charge involves a significant change of the shape of the well and the position of energy levels in this well, free-electron tunneling will be impossible and the transmission coefficient of the system will sharply decrease. Using molecular fragments of different structures as controlling groups, one may achieve structural rearrangement of the devices

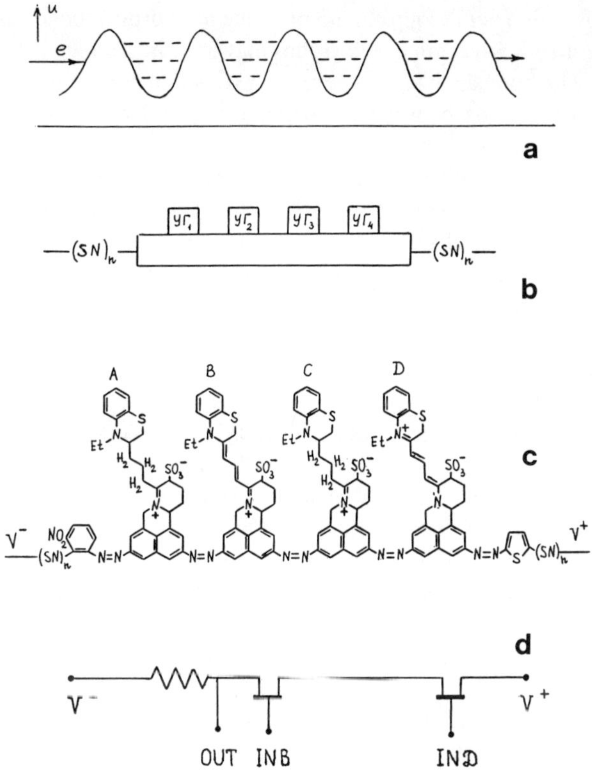

FIGURE 7. The molecular analog to a NAND gate (see explanation in the text).

employing different photon energies for different inputs. The described molecular element employs optical input of information. Other ways to influence the controlling groups are also possible, for example, by electrons[40] or by charged solitons.[40]

The second approach developed by Carter[5,6] is based on rearrangement of the electronic structure of the molecular system caused by collective electron-nuclear processes.

Among the processes of excitation energy transfer in pseudo-one-dimensional molecular chains, the soliton-like mechanisms appear to be of particular interest. The ideas of Davydov and collaborators[47,48] appeared fundamental in gaining a better insight into the soliton mechanism in molecular systems providing a number of biological events.

There are at least two types of nonlinear systems where solitons may propagate. The first of them is the molecular chain built of bonded fragments, weakly interacting, for instance, because of additional hydrogen bonds. As a result electronic and vibrational properties of the chain fragments slightly differ from those of individual molecules. An example of such a system is the chain of amino acid groups interacting due to hydrogen bonds in α-helical protein fragments. The amino acid groups may have intramolecular vibrational excitations with comparatively large transition dipole moments providing strong resonance interactions between the neighboring molecules and resulting in collective excitations of the exciton type.

Another kind of collective excitation of the soliton type in these molecular chains can result from mutual balance between dispersion (associated with resonance interaction of intramolecular vibrational excitations) and nonlinearity caused by the coupling of these vibrations with local displacements of fragments from equilibrium positions. Solitons represent stable local formations which propagate along the chain at rather long distances without appreciable energy dissipation.[47-48]

The second type of systems having inherent collective effects is conjugated polymers (for instance, *trans*-polyacetylene). The existence of soliton-like excitations of electron density in these systems is due to degeneration of the ground electronic state.[49] Displaying soliton excitations in conjugated polymers is believed[40,41] to be the basis for the functioning of various molecular devices. For instance, conjugated polymers may act as "molecular wires" in the process of signal transmission between molecular device fragments if the signal is a moving soliton excitation.

Soliton excitations were theoretically predicted from detailed analysis of the electronic structure of conjugated carbon-carbon bond systems of a rather long length. It can readily be seen that a system of conjugated bonds of the *trans*-$(CH)_n$ type having an odd number of carbon atoms may exist in two energetically degenerated forms, which may give rise to a defect in the electronic structure (Figure 8).

This problem was theoretically treated by Pople and Walsmley[50] in 1962. To overcome the difficulties arising from putting into consideration boundary conditions, the authors in Reference 50 examined a closed cyclic polyene chain where two defects may simultaneously arise; they were able to show that in addition to the valence and conduction bands, there appear two localized nonbonded orbitals corresponding to these two defects. Since these orbitals are very close to the valence band, their population (according to Pople and Walmsley's estimation[50]) is close to 1.4% at T = 300 K that corresponds to approximately one defect per 70 carbon atoms. Moreover, the same authors[50] predicted that these defects, i.e., electrons having parallel spins and occupying a localized orbital, should be manifested in electron-spin resonance (ESR) spectra. Indeed in the late 1970s, the ESR polyacetylene spectra studies confirmed the presence in a pure polymer of paramagnetic defects which increased in number with an increase in *trans*-form content.

Theoretical analysis[50] enables correct qualitative description of the arising structural defects. If an electron comes to a nonbonded orbital, the paramagnetic defect appears in the molecule (Figure 8). If the electron leaves the orbital or one more electron comes to the orbital, there arise positively or negatively charged defects similar to the well-known carbon ion (Figure 8).

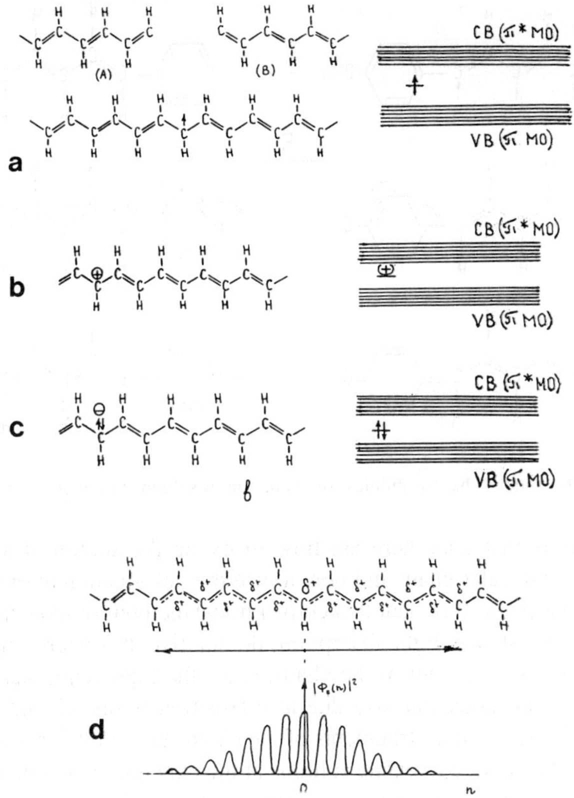

FIGURE 8. Diagrammatic representation of solitons in *trans*-polyacetylene.

A more exact theoretical consideration of the electronic structure of the polyacetylene *trans* form was done by Su et al.[51] They showed that the Shroedinger equation for the extended *trans*-polyacetylene chain has a soliton-like solution corresponding to the defects described by Pople and Walsmley.[50] It was also demonstrated[51] that both neutral and charged soliton-like states may arise when spin $S_o = {}^1/_2$ corresponds to neutral solitons and spin $S_{\pm} = 0$ to the charged ($Z_{\pm} = {}^{\pm}1$) ones. The average length of soliton excitation appears to be the same for neutral and charged solitons, and its halfwidth is about seven carbon-carbon bond lengths.

The probability of electron-density distribution inside the soliton has a pseudoperiodic character (Figure 8).

The authors concluded that the neutral soliton does exist in pure poly-acetylene. At the same time, they explained the conductivity of polyacetylene, doped with alkaline metals or halogens, by excitation of charged solitons.

It should also be noted that despite the progress in understanding soliton mechanisms underlying the transfer of energy excitations, electrons, and protons, it still remains unclear how these mechanisms can be implemented. The

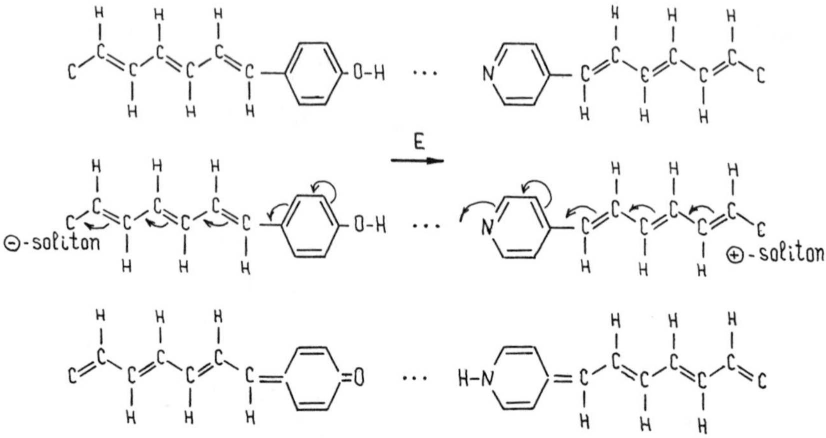

FIGURE 9. One of the possibilities for excitation of solitons in a molecular assembly.

main problems that arise here are how to excite the soliton at a particular point of the molecular chain and how to use this information after the soliton has passed the chain. It is clear that conventional methods of molecular system excitation (for instance, light absorption) do not give the opportunity to excite the soliton at a given point of the chain, since the light beam that cannot be focused down to molecular size due to diffraction limits will interact (with equal probability) with all fragments of the chain. Davydov,[47,48] Carter,[40] and Carter et al.[41] proposed to excite solitons at the expense of chemical reactions between specially chosen terminal groups of the chain. Carter[40] and Carter et al.[41] suggested that the proton transport from one molecule to another can induce excitation of both positively and negatively charged solitons according to the scheme in Figure 9.

It can easily be seen that soliton propagation in a pseudo-one-dimensional conjugated system is accompanied by replacement of double bonds by single ones, and vice versa. It can be used for designing a switch, incorporating into the original conjugated system an externally controlled molecular fragment capable of rearranging the chain of conjugation.

An example of such a fragment is the molecule of 1,1-*N,N*-dimethyl-2-nitroethenamine that, when incorporated into the chain of *trans*-polyacetylene, can absorb light and in the process of photoconversion is able to rearrange the conjugated chain (Figure 10). As a result:

1. After soliton passes through the main chain and there is replacement of double bonds by single ones (and vice versa), it will be impossible to excite photoisomerization of the control fragment that can be "an indicator" of soliton passing.

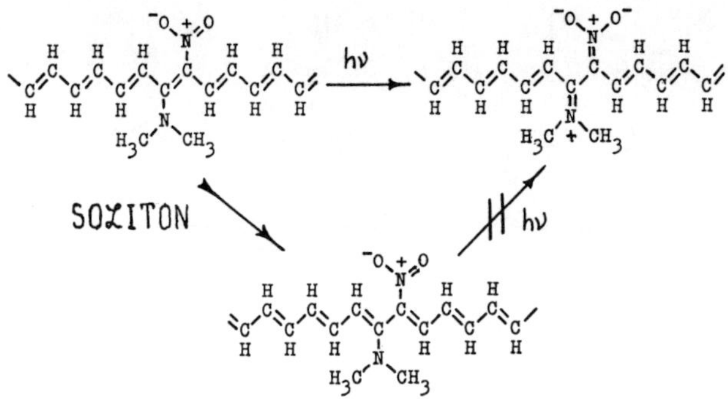

FIGURE 10. An example of soliton switching.

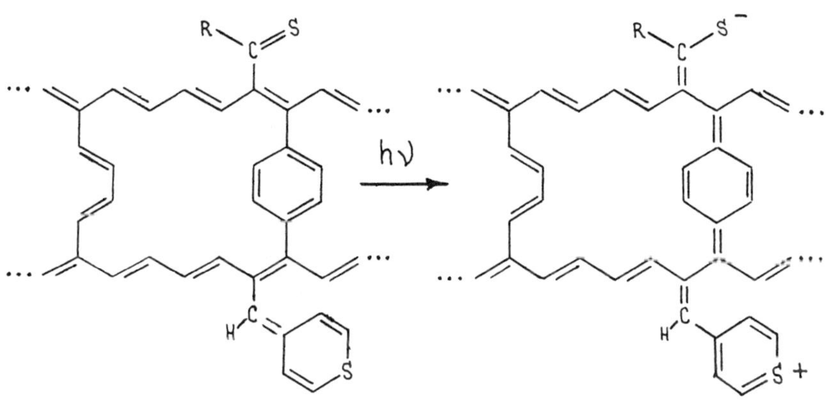

FIGURE 11. Soliton switching in a complicated molecular assembly.

2. Isomerization initiated before the soliton has passed the main chain will alter the electronic structure not only in the controllable fragment but also in the adjacent region, thereby breaking the conjugation chain and making soliton passing impossible.

The concept of soliton switching can be extended into systems containing several conjugated pseudo-one-dimensional fragments through which solitons can propagate. An example of this situation is given in Figure 11 showing two conjugated chains connected by two different chromophore groups. The structure of chain 1 is that chromophore 1 will not absorb light in the absorption band of chromophore 2 while chromophore 2 may appear photoexcited. As a result, soliton traveling along chain 1 will switch on chromophore 1 and switch off chromophore 2. Soliton moving along chain 2 will switch off both chromophores.

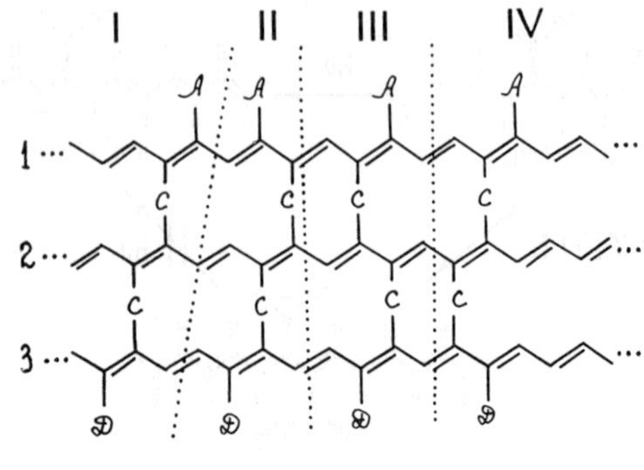

FIGURE 12. The soliton gang switch having several states (see explanation see in the text).

TABLE 2
Three-Chain Gang Switching

Chain	Channel I	II	III	IV
1	1	0	1	1
2	1	1	0	1
3	1	1	1	0

Note: A channel is only open if a vertical column is all ones. Soliton passage through a chain changes all ones to zeros and vice versa.

Similar principles may be used in designing more complicated switching systems having a great number of input and output channels. Figure 12 shows such a system built on the basis of electron-conducting chains A . . . C . . . C . . . D joined together with conjugated bonds. Here A, C, and D are an electron acceptor, intermediate group, and electron donor, respectively. Similar to the case of NAND gate considered by Carter[40] and Carter et al.[41] this system is based on resonance tunnel transition of the electron. Soliton switching of inputs (I–IV) can be realized using three channels built of polyene chains (horizontal directions in Figure 12). Note that each group (A, C, D), separated from one another by a dotted line can be in different states, depending on the distribution of single and double bonds. The possible states of the system as a whole are given in Table 2. Here the "1" stands for a double carbon-carbon bond between A and C, C and C, C and D in respective channels. It can easily be seen that an electron passing along the chain (shown in Figure 12 by the A . . . D vertical line) may occur only when there are just unities in the respective column of Table 1. For instance, fragment C is switched off from A at input II by chain I, while fragment C is switched off from the second fragment C at input III by chain 2. The state of the system will significantly change as the solitary wave propagates along the corresponding chain.

Carter[40] and Carter et al.[41] proposed a great number of other hypothetical molecular systems capable of performing different logical functions. Among them are the intersecting chains of *trans*-polyacetylene with two common carbon atoms, three semi-infinite δ-connected chains, soliton inverters, soliton memory elements, etc. The possible design solutions for simple logical circuits based on the elements mentioned above were considered by several authors, including the more detailed consideration by Groves.[52]

This approach, accomplished in the 1980s by Carter,[5,6,40,41] can be the basis for the reasonable system known in literature as soliton logic. Unfortunately, devoid of direct experimental support, the approach appears to be speculative. Moreover, it is evident that synthesis of molecular systems implementing Carter's elements as well as integration of these elements into a single device is a very complicated problem, the approaches to which are not sufficiently clear today.

However, trying to estimate the nature of Carter's conception it should be noted that it is in its essence an attempt to describe chemical transformations of conjugated organic molecules in terms of physical phenomena. This interpretation of Carter's conception allows one to outline new aspects of the problem of soliton logic.

As is known,[34,35] a possibility exists for synthesizing molecules containing both donor and acceptor fragments simultaneously. Bearing this in mind let us consider the molecules (supposing they are stable) of the type found in Figure 13.

Then, mentally arranging these molecules into pseudolinear chains (Figure 13) and proceeding from common chemical conceptions, it is possible to expect hydrogen bond forming between molecules and proton tunneling along this chain. Proton tunneling should be accompanied by rearrangement of the electronic structure of molecules that is formally equivalent to soliton passing (Figure 13).

The common features of such systems and some of the simplest analogs of the Carter elements are shown in Figure 13. However, proceeding from this chemical approach:

1. It becomes evident that one cannot consider the switching soliton-controlled systems without knowing where and how the soliton can arise and how it is transformed after passing along the chain (see also Roth[53] and Vanderbilt and Mele[54]).
2. If one takes into account the mechanism of soliton excitation shown in Figure 13 and its "departure" from the molecule chain, it can be concluded that soliton switching should be accompanied by proton transfer, substantial rearrangement of the molecular structure as a whole (not

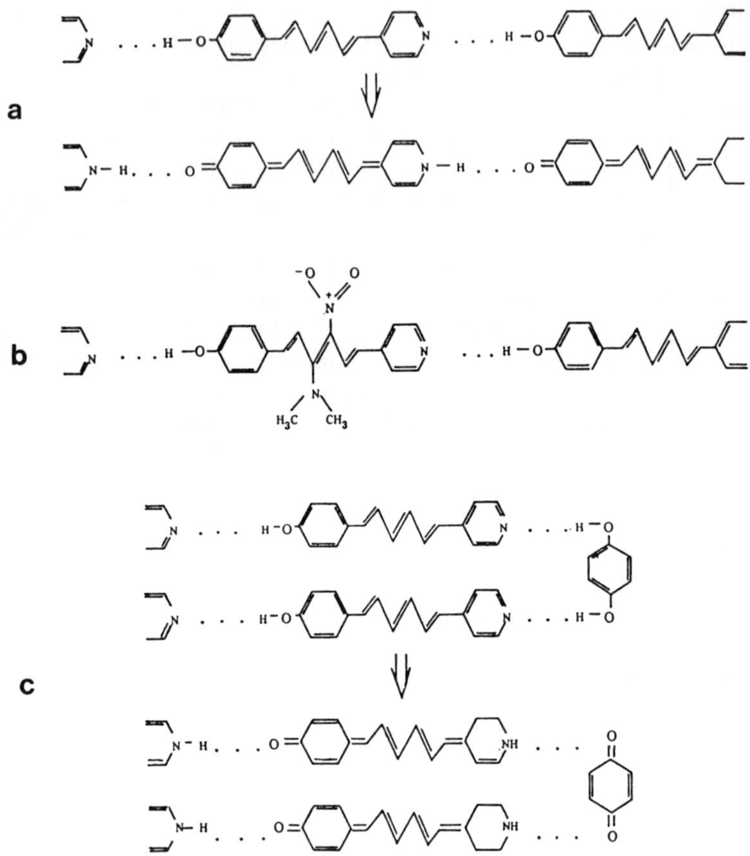

FIGURE 13. Chemical analogies to soliton switching.

only changes in the sequence of double and single bonds but rearrangements of the terminal fragments), redox reactions, etc.

3. It might be suggested that in order to avoid unnecessary "interconnections" (at the points where chemical reactions are to occur) it would be more efficient to synthesize most complicated molecular elements (such as summators, discriminators, etc.) controlled by chemical reactions triggered by physical factors.

The chemical approach discussed may be helpful in analyzing the possibilities of synthesizing the Carter molecular logical elements taking into consideration molecular stability, isomerization, dynamic molecular rearrangements, steric interactions of nonbonded groups, etc. Moreover, it can give an opportunity to transfer Carter's approach to the level of discussing the really existing organic compounds.

B. THE POTENTIALITIES FOR IMPLEMENTING COMPUTING DEVICES EMPLOYING MOLECULAR ELEMENTS

As mentioned above, numerous theoretical and experimental studies having the goal of elaboration of discrete molecular computing devices were investigated the possibilities of designing single molecular elements.

Even the most advanced theoretical investigations did not go further than consideration at the level of single logical blocks, for instance, Carter's attempts to design molecular circuits of the NAND and NOR type[40] or his proposal to elaborate cellular automatons using soliton switching circuits.[40,41] A radically different and new step in this direction was the work of Aviram[55] in which he developed a consistent theory of designing one of the possible versions of a discrete molecular computer.

As a matter of fact, the starting point of Aviram's proposal was the concept of electron tunneling through the system: metal — molecular electron acceptor fragment — nonconductive barrier — molecular electron donor fragment — metal[31] (see above). However, unlike Aviram and Ratner[31] the donor and acceptor fragments of the basic structure of a molecular computer in this model were two forms (conductive and nonconductive) of the same molecular system.

It is known (for detailed references see Aviram[55]) that there exist rather extended molecules (about 50 Å along the longest axis) that may occur both in conductive and nonconductive forms and pass from one form to another because of oxidation or reduction. The nonconductive form (Figure 14a) has the highest occupied molecular orbital having an energy less than the Fermi level of metal. Therefore, if such a molecule is placed between two metal electrodes to which low voltage is applied, no electron tunneling takes place. At the same time, the tunneling will take place if one electron is removed from the highest occupied molecular orbital, i.e., for the case of a positive molecular ion (Figure 14b). An example of such molecular systems is shown in Figure 14a (to the right).

As a possible primitive for the molecular computer Aviram proposed the molecule shown in Figure 14. In this molecule, two fragments (conducting and nonconducting) are coupled by a nonconducting hydrocarbon spiro system in a way that the longest axes of the fragments appear to be perpendicular to each other.

There are three features that are significant for the molecules under consideration. First, each of the fragments should be long enough to given an opportunity for the highest partially filled orbitals of the fragment to form a conductance band. In other words, these fragments should be analogs of the conductive polymers. After applying voltage to the ends of the fragment fixed on the metallic electrodes (Figure 15a), this allows us to determine in which state (conductive or nonconductive) it is.

Second, the structure of the molecule as a whole and of the nonconducting hydrocarbon bridge should be such that electron tunneling is carried out at a

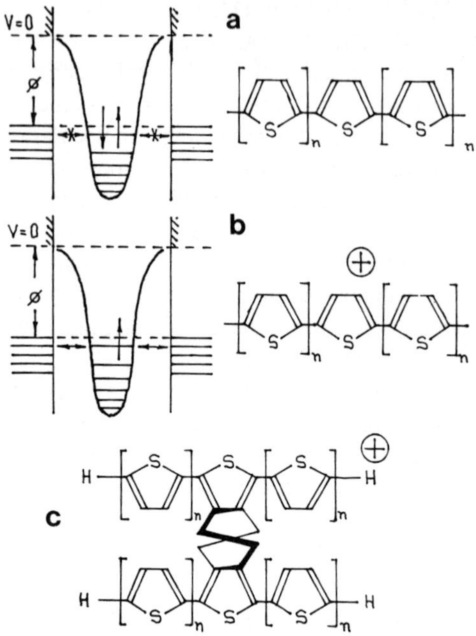

FIGURE 14. The relative arrangement of energy levels of a proconductor (a) and a conductor (b) molecules between two metallic electrodes and an illustrative example of a proconductor-σ conductor molecule (c).

relatively low threshold potential according to these mechanisms: metal-conducting system (which converts the conducting form into nonconducting) and metal-nonconducting system (which converts the nonconducting form into conducting).

Third, switching of the molecule, i.e., conversion of a conducting fragment to a nonconducting one, and vice versa, takes place under the action of the electric potential applied to two microelectrodes placed on the central axis of the molecule (Figure 15a).

Thus the molecular primitive proposed by Aviram is a field-controlled switch having two states. It should be noted that implementation of Aviram's model requires additional investigations to choose finally the most appropriate fragments of the molecular switch and to elaborate very complicated technologies of synthesizing molecular primitives and assembling them on a substrate into a planar system similar to the present-day integrated circuits.

The main logical blocks built on the basis of proposed primitives[55] are shown in Figure 15b. The arrows denote the passage of electrical current through one of the fragments. In the middle of the square (Figure 15a) the sign of the voltage on the upper control electrode is shown.

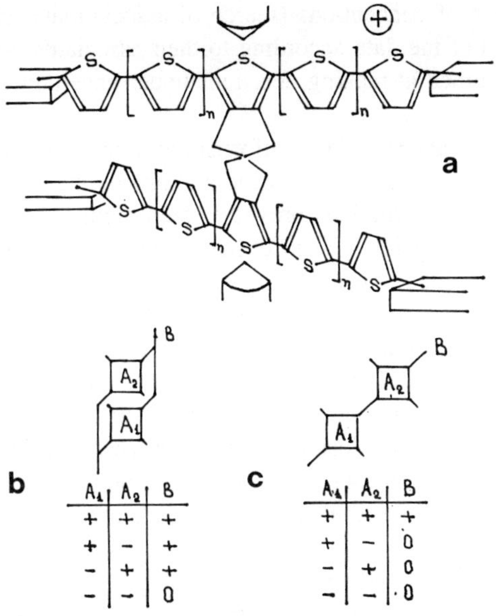

FIGURE 15. The simplified model of a molecular switch (a) and wiring diagrams for OR (b) and AND (c) gates based on this molecular switch.

In this review molecular and biomolecular computing and information processing devices are primarily discussed. However, mention should be made here on elaboration of molecular ROM having a memory capacity of several hundreds Gbytes[56] and 40 Mbytes fast optical RAM based on bacteriorhodopsin[57] (see also the helpful review article by Wild and Renn[58]).

III. BASIC APPROACHES TO THE IMPLEMENTATION OF NONDISCRETE MOLECULAR COMPUTING AND INFORMATION PROCESSING DEVICES

A. GENERAL PRINCIPLES OF INFORMATION PROCESSING AT THE MOLECULAR LEVEL

Information processing, i.e., transformation and transmission of the input data flow into the output one, has a wide variety of problems including:

- Reproduction of information
- Solution of different mathematical problems ranging from algebraic equations to nonlinear differential ones with arbitrary initial and boundary conditions
- Recognition processes including classification, image analysis, etc.

- Elaboration of conceptions (search of associations), classification and generation of the data according to their associative attributes
- Analysis aimed at making decisions under uncertain conditions, etc.

For over 40 years Shannon's information theory has been the paradigm in the field of information transmission and processing.

This system of conceptions proved to be a highly successful description of the operational principles for information transmission devices; therefore, it was widely accepted. Recently, however, in connection with progress in the study of information features of chemical, biological, and other complicated dynamic objects this formal description has appeared to be inadequate.[59] The main reason is that Shannon's theory defines comprehensively the quantity but not the quality (value) of information. At the same time, whenever changing the quantity of information in the process of calculations is examined, the measure of information meant is the one that reflects (apparently or not) the degree of its usefulness and depends on:

- What problem the information processing device solve
- What is known *a priori* before the beginning of solving the problem
- How the information generated by the device can be interpreted

The conception of valuable (useful) information satisfies these requirements (see, e.g., Chernavskii[60]).

Besides, one of the major problems in understanding the operational features of information processing systems is what the simplest correlations between hardware complexity (C_H) and software complexity (C_S) are. For electronic systems C_S is conventionally assumed to grow linearly, as the complexity of C_H increases. Recently, however, primitive mechanical, biological, and other systems demonstrated complicated behavior, despite the simplicity of their structures (for details see Nicolis[59]).

Finally, Shannon's theory does not contain such important conceptions as simulation, abstraction, and ability to memorize, which are typical not only of biological but also of some other systems having a simple structure.

Thus, recently there has arisen the belief that there should exist information processing systems having complicated behavior, which do not use the man-made principle of device complication to achieve a high behavior complexity.

One possible means of designing such systems is the use of basic principles of information processing at the molecular level that are materialized in primitive biological (supramolecular) structures (see Section I).

These principles, their display in the known biomolecular information processing systems, and corresponding molecular mechanisms were discussed in Rambidi.[61] Therefore, let us pay further attention to the possible ways of their implementation in biomolecular devices for information processing.

B. POSSIBLE WAYS OF ELABORATING BIOMOLECULAR COMPUTERS

1. Information Processing Features of Complicated Nonlinear Dynamic Systems

Reliable information processing is based on the existence of a code (language) in the system of data processing, i.e., a set of rules that generates information at a given hierarchical level of the system and compresses it at a higher cognitive level (for details see Nicolis[59]).

To achieve this goal, the language should meet the requirements of optimum relation between changeability that is determined by the stochasticity of the system and its capability of finding and correcting errors. The latter, in its turn, requires that the system have a memory. The dynamic systems, described at least by three first order nonlinear differential equations, satisfy these requirements.

As is known, the conventional method of characterizing a dynamic system is a study of the time evolution of a point in the state space. It is important to realize what is the character of final states that can be reached by the system in the process of its evolution. The principal types of these states are steady points and limit cycles. The state space closed domains, wherein the system moves in the stochastic dynamic mode, are conventionally named strange attractors.

A remarkable feature of nonlinear dynamic systems is that they appear to be a natural ground to implement information processing devices capable of solving complicated problems.

First, the multistable dynamic system having stable states is a system having memory. From an arbitrary point of phase space the system comes to an attractor, where it remains until an external action brings it out of this state.

Second, if the system is characterized by several attractors, the state space is divided into domains (attractor basins), wherein the point representing the system comes to its attractor. In such a way the dynamic system simulates a process of abstraction, since a set of the states from each basin is characterized by one "collective" element.

Third, a change of the boundary and initial conditions as well as constants of the system being described results in a change of a set of attractors. If it is possible to set feedbacks allowing the system state to be compared with a previous one, then the dynamic system will be capable of controlled evolution (self-learning).

Fourth, using the attractors of some system as an input information for another dynamic system, a multilevel information processing system may be implemented. In such a system at a cognitive level the attractors will mean decision making.

In recent years detailed investigations have been conducted on biological (biomolecular) objects that have demonstrated at different structural (morphological) levels complicated nonlinear processes. Among them (in the succession of decreasing structural complexity) are

- Neuronal nets activity and evolution of stochastic waves in the myocardium (body tissue level)[62,63]
- Dynamics of biological population evolution (cell level)[64]
- Biochemical reactions in cells and membranes (supramolecular system level)[64]
- Collective excitations in biomacromolecules (molecular level)[10,48]

In such systems there may exist modes as follows:

- Stationary nonhomogeneous distributions of the system components (spatial structure)
- Oscillatory or pseudooscillatory regime (temporal structure)
- Regular autooscillation phenomena (spatiotemporal structure)
- Randon temporal or spatiotemporal behavior (turbulence)

Among different structural levels the supramolecular systems may be the most suitable for the implementation of information processing devices. The behavior of these systems is much more complicated than that of single biomacromolecules. At the same time the specific ways of forming supramolecular systems are determined by the possibilities of laboratory technology rather than by the methods of extracting and keeping alive bacterial cultures or tissues. Besides, almost all the previously mentioned general principles of information processing at a molecular level are realizable on the basis of supramolecular systems.

Therefore, let us imply by the nondiscrete biomolecular computer a multilevel system; each operational level of the system is a supramolecular structure formed on the basis of macromolecules. The behavior of the system is described by the complicated nonlinear dynamics. In this system:

1. An information exchange between levels should occur, i.e., the attractors of processes at the low-lying level should be the input information for the following level.
2. The structure of the system and its levels should be optimum for the highly parallel information processing.
3. The system should be capable of self-learning.
4. Generation of valuable information should be provided.

Different versions of implementing the supramolecular structures having complicated nonlinear dynamics have been considered.[10,23,64-67] These versions are based on different physicochemical mechanisms of the dynamic processes: mechanochemical interactions,[10,23] collective nature of processes in "artificial biomembrane-protein channel" systems,[65] etc. The biochemical systems based on nonlinear kinetics of the chemical reactions together with diffusion of reaction components[64,66,67] belong to a more general type of biomolecular systems showing diverse types of behavior.

2. Design Principles and Possible Operations of Biomolecular Computers

Given the information properties of nonlinear dynamic biomolecular systems, one may picture some versions of the information processing devices that may be embodied practically. Their structure defines two general type design solutions, differing in that the levels of information processing may be spatially combined or spatially separated. In spatially combined systems the successive levels (stages) of information processing are materialized as a set of consecutive dynamic processes (chemical reactions) that proceed in the same volume, the reaction products of some levels being reaction substrates for the following levels. In the spatially separated dynamic devices the processes of different levels are localized in certain domains of supramolecular structure. For instance, a device of this type may be formed based on a system of the parallel-operating pseudoflat primitive elements, each of them (or several of them) corresponding to a certain level of information processing (Figure 16). A single primitive element should be a limited size flat carrier (artificial biomembrane, polymer film, etc.) that contains immobilized individual components of the biochemical system (enzymes, receptors, and other biomacromolecules) used. Optimally, the carrier should be porous to provide transport of the reaction components through the device, i.e., energy supply. The interconnections between different levels may be of the chemical type, i.e., the products of reactions at some levels being substrates for reactions at the next level. A flat two-dimensional structure of elements allows parallel information input-output to be easily realized and parallel information processing to be carried out (for instance, input of information by projecting the light pattern onto the active layer that transforms the light signal into spatial distribution of the photochemical reaction product).

Given the nondiscrete character of information treatment in the process of biomolecular computer operation there is a temptation to call this type of device analogous. Nevertheless, there is a fundamental distinction between analogous and biomolecular devices. The analog evaluator, for instance, is a dynamic system that can be investigated and that is functionally homomorphous to another dynamic system whose characteristics cannot for some reason be determined. Unlike the analogous devices, in a biomolecular computer it is possible to pick out single primitive elements that fulfill definite primitive operations and are characterized by definite transfer functions. The solutions of problems under investigation can be achieved on the basis of these operations (functionally complicated in the case of a biomolecular computer). Thus the treatment of information by the nondiscrete biomolecular

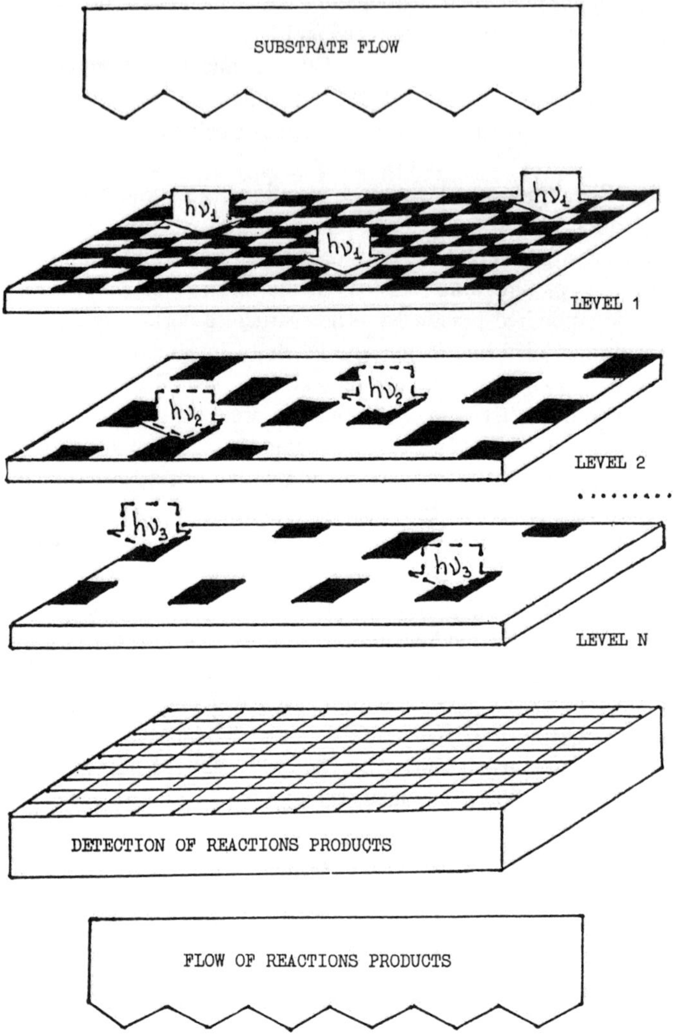

FIGURE 16. The simplified model of the multilevel biomolecular information processing device. Arrows represent optical control (input of information) in different frequency ranges.

device is performed based on the same main principles as in digital devices but without transforming information into a discrete form.

Given the principal ideas of biomolecular information processing devices elaboration one may suggest a number of basic operations that can be fulfilled by the biomolecular computer, metaprocedures for realization of these operations, and possible models of supramolecular systems by the use of which these metaprocedures can be materialized.

Some operations and possible ways of their implementation have been considered in References 10, 23, and 68–70. For a fuller list of the operations see Table 3.

The principles of information processing at the molecular level being the basis for the nondiscrete biomolecular devices (see above) permit more complicated information processing operations than those presented in Table 3. Especially, the possibility of the use of these principles for simulating such fundamental modes of nervous system activity as ''learning'', ''keeping attention'', and ''intermixing layer'' should be highly promising.

Learning is one of the versions of programming biomolecular information processing devices. Unlike conventional digital computers, in the process of molecular device learning there occur real changes in the nature of interconnections between elements and hence changes in the structure of devices. One of the basic features of nondiscrete devices is the manifestation of interdependence and intersupplementability in ''hardware-software'' systems.

The concept of ''learning'' is rather vague and determined by the type of proceeding processes and by the complexity of realizable macrooperations.

In the case of pattern recognition, a set of reference patterns are presented to the device that results in a change of the device structure in a way that leads to the capability of fast recognition of the pattern from the submitted series even though the pattern is noisy.

Among more complicated versions is learning based on selective memory. In this case the system proves to be capable of remembering only the patterns close to one of several ''reference'' patterns being in its ''nonexplicated submemory'' (analog of the human subconscious). Learning based on selective memory may find use in designing self-learning systems for long-term monitoring.

Supplementary to the selective memory is the learning using a ''novelty filter'' that changes the device structure only if the presented pattern essentially differs from those stored in the memory. The presence of the ''novelty filter'' markedly improves the device information performance, e.g., prevents one pattern from ousting all the others from the device memory.

The selective learning and ''novelty filter'' realizable within the framework of the neural net paradigm have been considered in References 71, 72 and 73, 74, respectively.

The learning principle is rather difficult to realize technically. Biomolecular processors having high behavioral possibilities are likely to solve this problem more effectively than in the case of conventional computers.

The keeping attention principle means that a computer or processor (or its part) may be originally in one of two states: active (attention) and passive. Operation efficiency in the active state is higher than that in the passive one. However, the reliability of the system as well as its protection against external random factors is higher in the passive state than in the active one. Thus, the system operation efficiency and the reliability of the system are in the broad

TABLE 3
Major Operations of Nondiscrete Biomolecular Computers, Metaprocedures Embodying These Operations, and Implementation of These Metaprocedures Using Appropriate Physical Models

Operations	Metaprocedures	Physical models for the realization of metaprocedures
Parallel arithmetic	Addition of data arrays	Planar system of photoactivated enzymes that produce local concentration of reaction products proportional to the total intensity of the light images simultaneously projected at the surface of the system
	Modulo 2 addition of data arrays	Planar two-level system; at the first level two types of photoactivated enzymes are immobilized; first of them implements summation of light images, the second changes of the acidity of the medium proportional to the intensity of light; products of the first level enzymes are substrates for enzymes of the second level, whose activity is controlled by the acidity of the medium
	Multiplication of data arrays	Planar two-level system; at the first level two different types of enzymes photoactivated in different frequency ranges are immobilized; products of these enzymes are substrates for the enzyme of the second level, which produce local concentration of the reaction product proportional to the product of the intensities of two light images projected at different frequencies at the surface of the first level structure
	Calculation of reciprocal values in data arrays	Planar two-level system; Photoactivated enzymes of the first level change the local acidity of the medium proportional to the intensity of projected image; activity of the second level enzymes decreases proportionally to increasing acidity of the medium
	Averaging of data arrays	Consecutive accumulation of light images by photosensitive media, for instance, by the bacteriorhodopsin containing films
Solution of complicated nonlinear problems	Investigation of systems implementing nonlinear problems under complicated (arbitrary) initial and boundary conditions	Molecular systems having complicated nonlinear dynamics
Recognition	Preliminary treatment	
	Filtering random noises	Homogeneous chemical oscillatory systems, biomolecular distributed multibox systems functioning in the dissipative pattern formation mode

TABLE 3 (continued)
Major Operations of Nondiscrete Biomolecular Computers, Metaprocedures Embodying These Operations, and Implementation of These Metaprocedures Using Appropriate Physical Models

Operations	Metaprocedures	Physical models for the realization of metaprocedures
	Segmentation (decomposition of an image into simpler parts)	Planar molecular multibox threshold devices having low spatial resolution, biomolecular distributed multibox systems functioning in the dissipative pattern formation mode
	Determination of the type of an image (triangle, square, circle, etc.)	Homogeneous chemical dynamic systems functioning in the traveling pulse mode
	Determination of the most important features of an image	
	Images sharpening (determination of the brightest parts)	Planar molecular multibox threshold devices having high spatial resolution
	Determination of the contour and angular points of an image	Homogeneous chemical dynamic systems functioning in the strata formation mode
	Determination of the connection	Homogeneous chemical dynamic systems functioning in the trigger mode
	Determination of the medial axes	Planar systems of photoactivated enzymes located at the biosensors assembly
	Categorization of images features	
	Estimation of the ratio of bright areas to dark ones	Planar system of photoactivated enzymes located at the biosensors assembly
	Calculation of figure areas, perimeters, centers of gravity, etc.	Homogeneous chemical dynamic systems functioning in the traveling front of switching mode
Search of the best solution of problems described by complicated tree diagrams	Search in the maze of possibilities	Chemical dynamic systems based on photoactivating and photoinhibiting enzymes that implement the pseudoplanar mazes corresponding to the essence of the problem under investigation
	Constructing of labyrinths of possibilities	Manufacturing labyrinths due to inhibition of preassigned parts of chemical dynamic media by a physical or chemical stimulus
Decision making	Decision making under uncertain conditions	Molecular and biomolecular systems implementing complicated nonlinear dynamics
	Description of the developing systems behavior	Supramolecular developing systems whose evolution is based on the principles of morphogenesis

sense supplementary (in the sense of Bohr) to one another, which may be related not only to computers but also to other complex systems including the living ones.

In living nature this dualism is resolved due to a switching from the activity mode to the rest one; this does not mean a complete cessation of all processes (death), since in the latter case reverse switching proves to be impossible.

The keeping attention principle is realizable in computers due to switching the whole system parameters, the measure of this change being of utmost importance. For instance, if one fully switches off the computer (or processor) energy supply to get the "rest" state, then the following switching-on takes too much time, often requiring readjustment of the device. What the optimum rest mode is and how it can be implemented are the issues presently under consideration.

The "active-passive" state alternative does not exhaust all aspects of the keeping attention problem. Particularly, the biomolecular device in some cases is capable of distinguishing patterns of a certain class without responding to all the others. Depending on the recognition problem, there exists a possibility of switching at each moment from one recognition class to another. A further complicated modification of the keeping attention principle is the context-dependent attention when the character of the extracted information is determined by patterns previously presented and recognized.

Specific examples of devices with such versions of the keeping attention principle will be discussed in the subsequent publications. This principle is not related to a certain class of problems; it is general and may be used in designing conventional numerical computers. However, in molecular devices it seems to be implemented in the simplest way.

The "intermixing layer" principle is of importance for solving some special problems.

Many important processes contain both stochastic and dynamic elements, their relative contribution being time dependent. Roulette and Chinese billiards give examples of such time dependence. In these cases the degree of stochasticity first increases (when the ball collides with reflectors), and then decreases to zero (when the ball penetrates into the region of attraction of a certain final state). In the state space of such a system one may single out a special domain, i.e., the intermixing layer that resembles (but is not in the exact sense) a strange attractor, since phase trajectories leave this layer.

The intermixing layer is required for solving problems including the stage of new valuable information generation, e.g., problems of decision making under uncertain conditions, evolution under competition, description of the human creative processes, etc.

In all the problems mentioned above the answer is of stochastic nature: its reliability increases depending on time. However, at the last stages of the solution the answer becomes trivial. Thus, the problem is reduced to finding a moment of the intermediate state, when the answer is rather certain but not yet trivial.

C. PROGRAMMING NONDISCRETE BIOMOLECULAR DEVICES AND OUTLOOKS

One more fundamental problem plays a principal role in understanding the essence of nondiscrete biomolecular computing. This is the problem of programming nondiscrete devices, which is currently the least elaborated one. The peculiarity of this problem is that the increase in possibilities for control of information processing (programming) is determined by the complication of the device structure. This feature virtually determines the outlooks for further development of nondiscrete information processing devices.

So far the theory of nondiscrete biomolecular systems for information processing has not been developed. Therefore, only principles of implementing separate elements but not the devices as a whole have been discussed here. These primitive elements should perform complicated operations of information processing at a certain level in the multilevel nondiscrete system or may be used as processors for digital computers. The possibilities of control of element behavior in this case are limited and determined by the degree of understanding the dynamics of the process that proceeds in the element and in the long run by the "design" of the element. Multilevel devices provide a broader variety of control possibilities.

Also discussed earlier was a primitive "one-channel" system, wherein single planar elements were formed in a "column" structure, and consecutive linear information processing was preset by the column structure of elements (Figure 16).

It is easy to understand that there are different possibilities of complicating the structure of the device and as a consequence their information processing possibilities. First, the levels of the column structure, which differ with respect to their position in a sequence, may be made the controllers. For example, some level may change the acidity of the medium under the action of the controlling physical agent and regulate the activity of enzymes of the subsequent levels. Second, a "multichannel" device may be designed on the basis of column structures by the confluence and the splitting of the channels (Figure 17).

The remaining question is whether it is possible to design a general purpose computer based on these primitive elements. However, it is clear now that complication of the structure of a nondiscrete device undoubtedly results in a considerable widening of its information processing possibilities.

The optimum system of simple logical operations of the digital device allows one to create a powerful programming system. That is why the process of solving the problem and cost of solution are determined by software (the cost of software may be about 90% of the total cost of the information processing system).

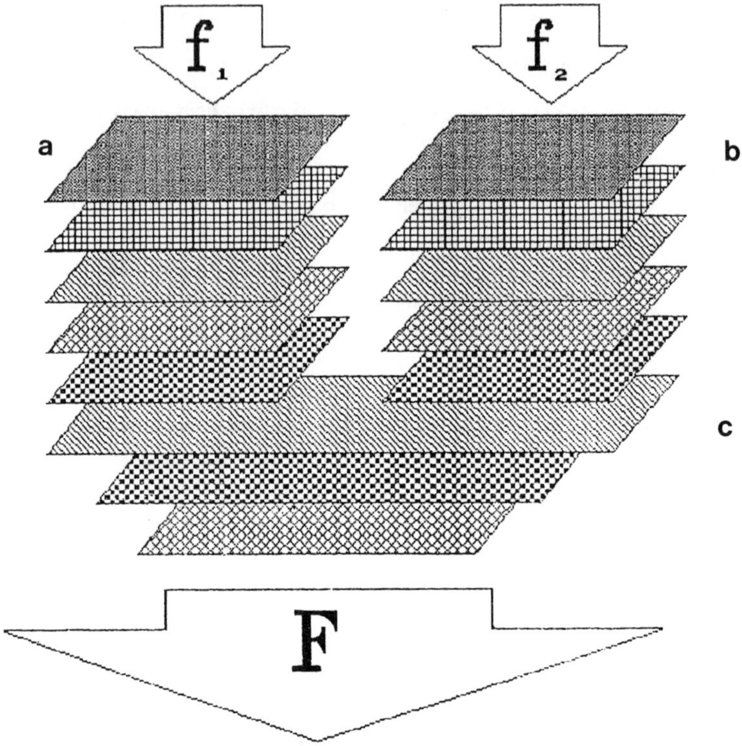

FIGURE 17. The simplified model of the multichannel biomolecular information processing device: (a), (b) "column" structures proceeding with information processing in channels 1 and 2; (c) an element that enables the confluence of two channels (for instance, nonlinear dynamic media functioning in the dissipative pattern formation mode).

For a nondiscrete biomolecular device the information processing possibilities are determined by the complexity of its structure. Therefore in this case, the designer performs a larger amount of work than the programmer, who puts necessary computational and information processing possibilities into the structure and dynamics of a device.

At the same time the high performance of nondiscrete primitive elements (as compared with the digital one) predetermines the possibility of solving problems of high computational complexity.

IV. INFORMATION PROCESSING AND COMPUTING DEVICES BASED ON BIOCHEMICAL NONLINEAR DYNAMIC SYSTEMS

A. KEY ELEMENTS OF BIOMOLECULAR COMPUTERS

To clarify the following discussion let us consider some terms.

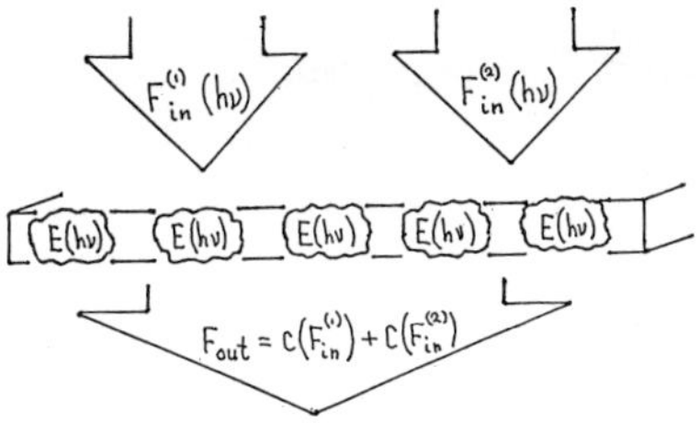

FIGURE 18. The simplified model of the biomolecular device implementing summation of the light images simultaneously projected on the input surface of the device. $E(h\nu)$ are the photoactivated enzymes.

Among basic elements of computing devices that materialize particular levels of dynamics, one should consider input and output information and logical blocks. The simplest example of an input pattern transducer is shown in Figure 18. It is seen that the photoactivated enzymes are immobilized in a planar supramolecular structure (artificial biomembrane, etc.). The light signal that activates enzymes causes spatial distribution of the reaction products in the membrane environment. It is clear that the transfer function "light intensity or light concentration", in this case, must be linear. It can readily be seen that elements of this type can be used to perform simple arithmetic highly parallel operations (e.g., summation of the elements of two optical images).

Logical blocks should combine potentialities to realize nonlinear information processing operations and the ability to solve complicated logical problems. Logical blocks performing operations of high computational complexity may also be used as special-purpose biomolecular processors incorporated into digital semiconductor-based computers.

An important point in construction of biomolecular devices is interconnections between single processing elements of each level of dynamics as well as between different dynamic levels. Biomolecular systems offer the possibility of organizing a nontraditional network of interconnections, for example, based on individual "information carriers" having different degrees of information transfer selectivity. The basic types of such interconnections as compared to those traditionally used are shown in Table 4.

B. STRUCTURAL AND DYNAMIC FEATURES OF BIOMOLECULAR DEVICES AND MODES OF THEIR FUNCTIONING

Nonlinear biochemical processes in supramolecular systems may form the basis of designing information processing devices performing operations of various computational complexity. Given an increasing behavioral complexity, e.g., the consecutively complicated dynamics and rising structural complexity, it is appropriate to consider the information processing capabilities of these devices.

TABLE 4
Interconnections in Different Information Processing Systems

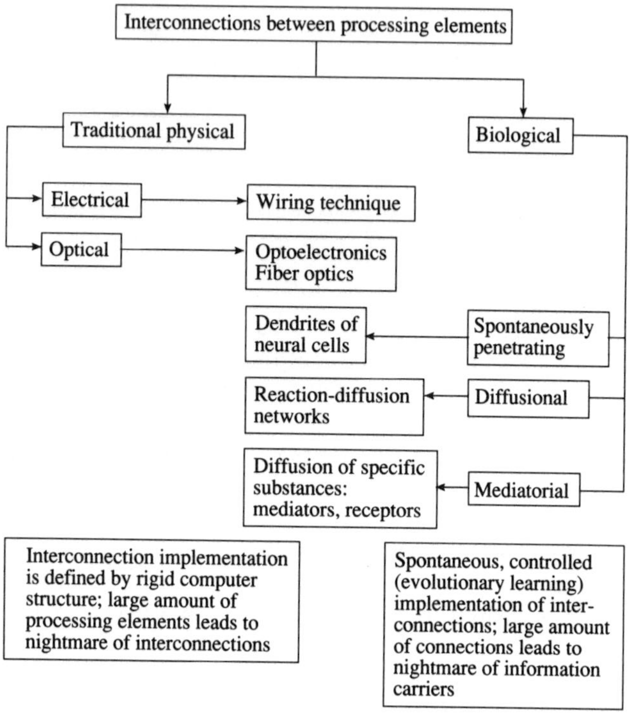

The simplest versions of these devices will be those employing one-level dynamic homogeneous systems and different complexity of reaction kinetics. The next step of complication is one-level homogeneous distributed systems with properties determined not only by the type of reaction kinetics but also by diffusion of the molecular components of the system. Finally, the most complicated ones are the distributed multilevel homogeneous systems showing high complexity of behavior.

A hierarchy of the behavioral complexity of devices becomes still more multiform if the possible structural differences of supramolecular systems are taken into account. In general, it is sufficient to consider two classes of systems differing in their structural characteristics. These are the spatially homogeneous supramolecular systems and spatially heterogeneous ones. Considering various combinations of dynamic and structural characteristics it appears possible to describe different blocks of the biomolecular computer (Tables 5–7).

Let us start with homogeneous supramolecular systems showing so-called "pointwise" kinetics (Table 5).

TABLE 5
Features of Spatially Uniform Dynamic Systems

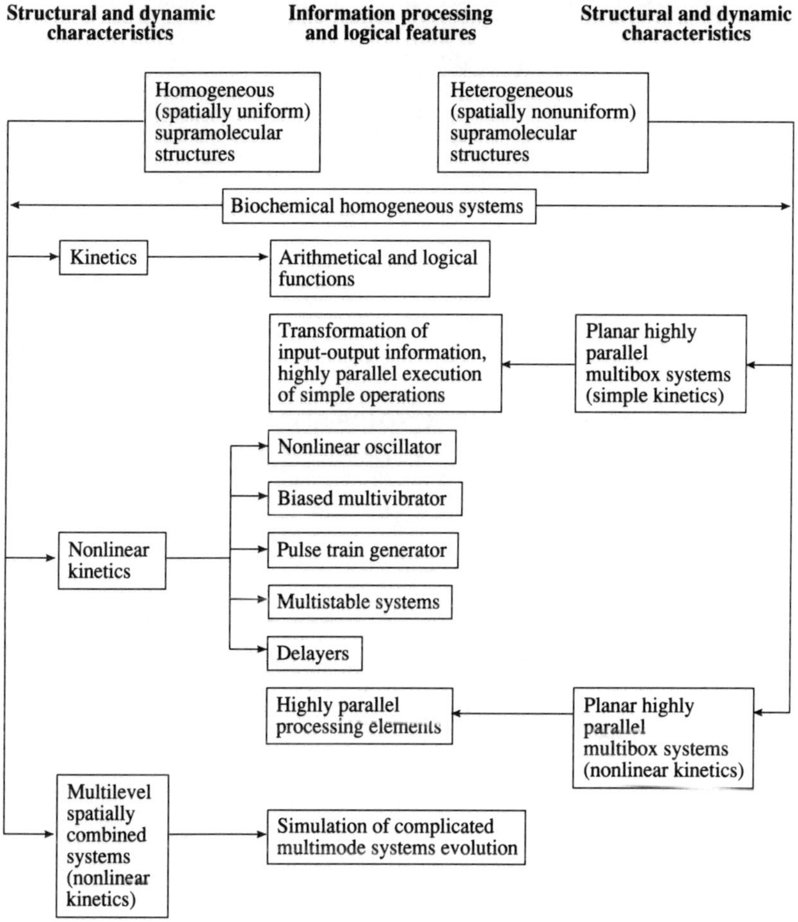

A trivial example of the system with simple kinetics can be the final volume of solution containing substances undergoing photoinduced chemical transformation under light radiation. It is possible to see that the system can be used as a nondiscrete summarizer of two simultaneously applied input light signals. If such a system has threshold characteristics, it is capable of classifying light signals into two groups according to the intensity value (below or above the threshold).

Perhaps greater possibilities are offered by homogeneous systems displaying complex nonlinear kinetics of reactions. A classical example is the Belousov-Zhabotinsky (BZ) reaction. Let us consider a simplified approximation to the BZ reaction, namely, the Field, Koros, Noyes model (see, for instance, Reference 67). In this approximation BZ reaction is described by the equations:

$$
\begin{aligned}
A + W &\xrightarrow{k_1} U + P \\
U + W &\xrightarrow{k_2} 2P \\
A + U &\xrightarrow{k_3} 2U + 2V \\
2U &\xrightarrow{k_4} A + P \\
B + V &\xrightarrow{k_5} hW
\end{aligned} \tag{1}
$$

where $A = [BrO_3^-]$, $B = [HOOC\text{–}CHBr\text{–}COOH]$, $P = [HOBr]$, $U = [HBrO_2]$, $V = [Ce^{4+}]$, $W = [Br^-]$. The kinetic equations describing this system take the form:

$$
\begin{aligned}
\epsilon\dot{u} &= aw - uw + u - u^2 \\
\dot{v} &= bu - v \\
\epsilon'\dot{w} &= -aw - uw + v
\end{aligned} \tag{2}
$$

where dimensionless variables and constants are such that:

$$
\begin{aligned}
&\epsilon' \ll \epsilon \ll 1 \\
&a \ll 1 \\
&b \approx 1
\end{aligned} \tag{3}
$$

Given these conditions and Tikhonov's theorem,[64] it is easy to arrive at the simplified two-component version of the Field, Koros, Noyes model:

$$
\begin{aligned}
\dot{u} &= u(1 - u) - v\frac{(u - a)}{(u + a)} \\
\epsilon^{-1}\dot{v} &= bu - v
\end{aligned} \tag{4}
$$

The main isoclines and time dependencies of system behavior corresponding to these equations are shown in Figure 19a, b, c. There they describe two types of behavior: steady stationary states (a and c [waiting mode]) and synchronous self-oscillations (b).

TABLE 6
Features of Distributed Biochemical Dynamic Systems

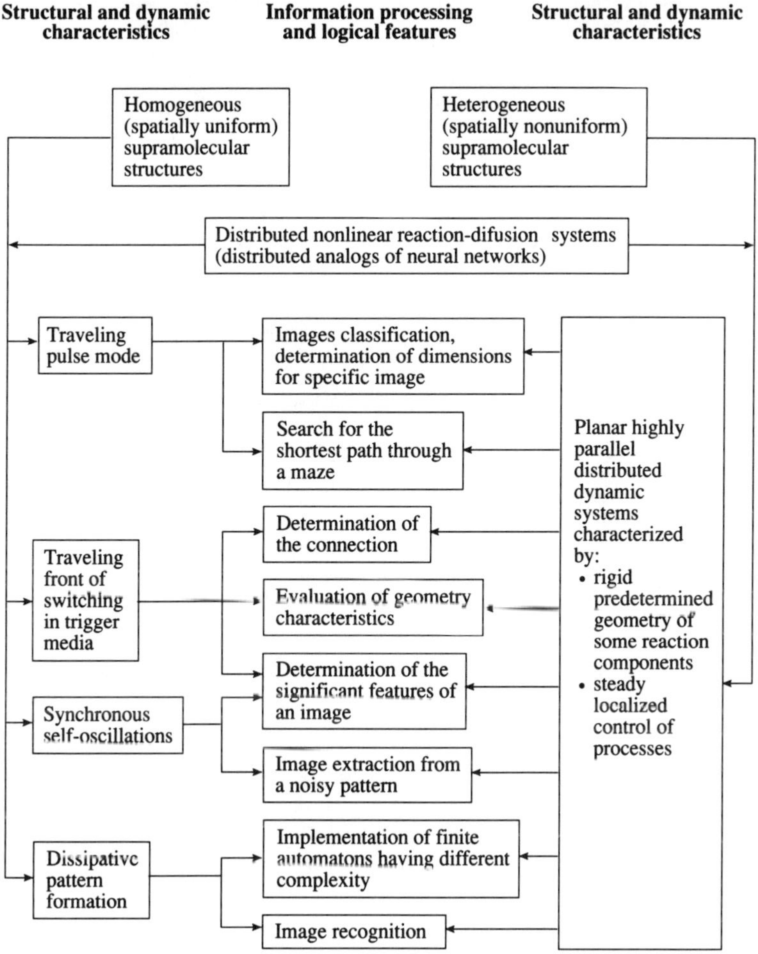

Aside from the BZ reaction it should be noted that other homogeneous systems employing nonlinear kinetics can be run in the mode when the system has two steady states (bistable system). The main isoclines and time dependencies of the system behavior corresponding to this case are presented in Figure 19(a).

These examples of homogeneous systems (pointwise kinetics) represent a description of physicochemical mechanisms rather than of devices to be built. Nevertheless, the practical significance of these mechanisms markedly increases when passing to heterogeneous parallel multibox structures. In this case (Figure 20) the device comprises a planar structure (for instance, an

TABLE 7
Features of Multilevel Heterogeneous Dynamic Systems

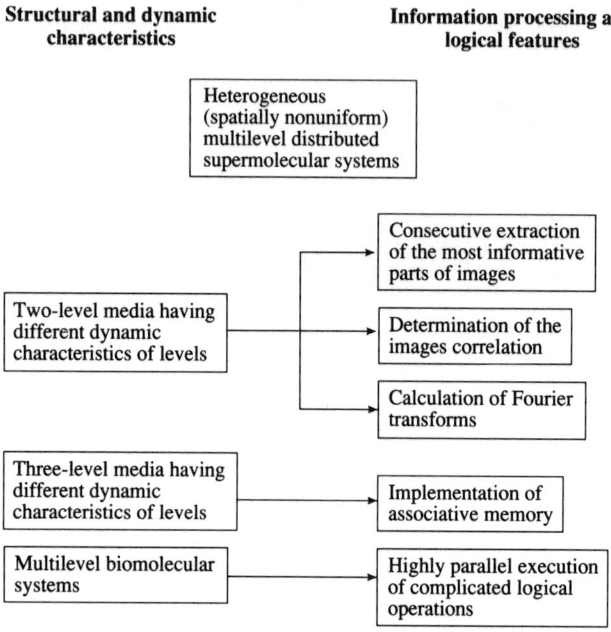

artificial biomembrane) divided into separate noninteracting independently functioning compartments (boxes) each operating by one of the mechanisms discussed. As can be seen, this allows building functional blocks that perform highly parallel arithmetic operations (Figure 18), determination of the significant features of images, etc.

The distributed biological dynamic systems combining nonlinear kinetics of reactions and diffusion of the reaction components can be run in different modes such as:

- Traveling pulse mode
- Traveling front of switching in a trigger medium
- Synchronous self-oscillations
- Dissipative pattern formation

In the last few years the possibility of employing these regimes for solving problems of high computational complexity has been examined by various research groups.[68-70] Perhaps one of the most significant information processing features of these nonlinear dynamic systems (otherwise excitable chemical media) is that they represent distributed analogs of neural networks. They provide a high degree of parallelism and, as a consequence, a high speed of information processing.

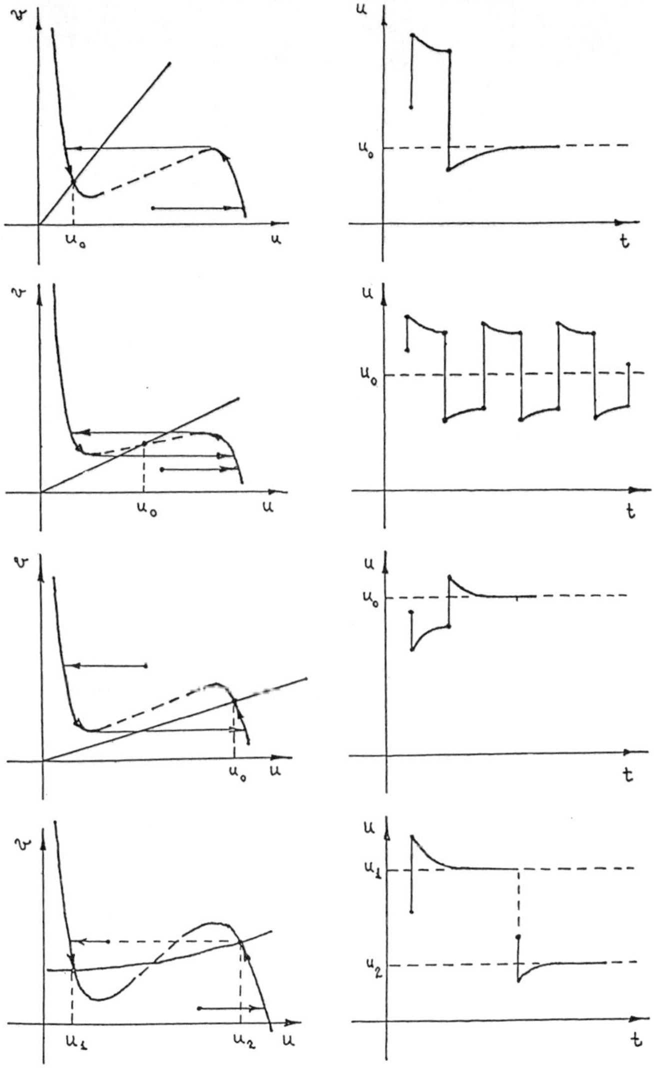

FIGURE 19. Main isoclines for homogeneous biochemical nonlinear dynamic system (see explanation in the text).

At the same time chemical excitable media appear to be poorer in their properties as compared to the known Hopfield model. Indeed, in chemical excitable media, for instance, the element of the medium is connected with several nearest neighbors only in contrast to neuronal networks where the elements of the networks are highly (fully) interconnected. Nevertheless, these seemingly simplified analogs of neuronal networks offer the possibility of obtaining significant results, (for instance, in pattern recognition; and of

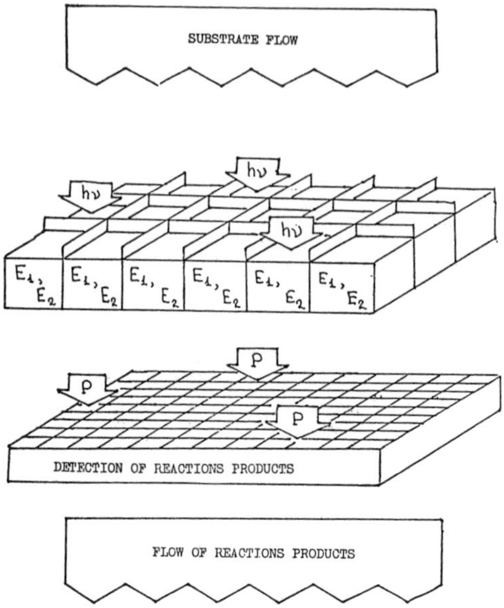

FIGURE 20. The simplified model of a biochemical multibox information processing device.

providing invariance of recognition with respect to translations, rotations, stretchings, and deformations of elements).[70]

An interesting possibility for implementation of biomolecular information processing devices is provided by the heterogeneous distributed biochemical systems that can be fabricated, for instance, using multibox compartmentalized supramolecular structures where the walls between compartments are permeable for molecular components (Figure 20). They provide a rigid preassigned geometry of the system components and the possibility of arranging steady local control. Let us consider some information processing features of the heterogeneous multibox systems (for more detail, see Sidelnikov et al.[75]).

The model is based on the reaction of oxidation of uric acid catalyzed by uricase enzyme:

$$\text{uric acid} \xrightarrow[\text{uricase}]{O_2,H_2O} \text{allantoin} \tag{5}$$

The reaction can proceed in the membrane with an immobilized enzyme. In this system where the membrane separates two solutions with a constant substrate and an excess of cosubstrate concentrations and where the reaction is sustained by diffusion of the substrate and cosubstrate from solution, the phenomenon of hysteresis was experimentally observed.[76]

Computer simulation of the dynamics in a single cell (one box) revealed that the system may have three steady states that can be conventionally termed as +1, 0, and −1.

Given the planar two-dimensional multibox system fabricated from uricase boxes and given a definite choice of model parameters, the dynamics of the system can be described by the equations:

$$\dot{x}_i = \delta x_i - x_i^3 - y_i$$
$$\dot{y}_i = \gamma x_i - y_i + d\left(\sum_{y_k \subset R_i} y_k - 4y_i\right) \quad (6)$$

where x_i and y_i are concentrations of substrate and cosubstrate in the ith compartment; R_i is the neighborhood of the compartment i; and δ, γ, i, and d are parameters of the model (their meanings are obvious).

Computer simulation of the system dynamics has shown that the homogeneous state of the system dynamics has shown that the homogeneous state of the system determined by uniform distribution of substrate and cosubstrate concentrations appears to be unstable with respect to local perturbations, i.e., to the increase of the concentration of the substrate (cosubstrate) at any point. In this case a dissipative pattern formation has been observed around the perturbation points. The structure of the pattern resembles a chessboard picture where the compartments with +1 and −1 states alternate (Figure 21). Depending on whether an even or odd number of compartments separate the perturbed boxes the surface is either completely filled with the chessboard pattern or dislocations of O-type states will appear at the interfaces (Figure 21).

Computer simulation has demonstrated that the dynamic processes proceeding in the system under consideration can be used as the basis to perform various complicated logical operations such as image extraction from a noisy pattern (Figure 22), extraction of the figure contour (Figure 23), smoothing the contour of a complicated image (Figure 24), etc.[27,75] More complete results of this computer simulation including the prospects of applying heterogeneous biochemical dynamic systems for fabrication of biomolecular information processing devices are being published in Gritsenko et al.[27] and Sidelnikov et al.[75]

Heterogeneous multibox biochemical systems can be used for implementation of multilevel devices where the products of reactions at any (particular) level will be substrates for reactions at the next levels (Figure 16).

C. SOME DESIGN CHARACTERISTICS AND OPERATIONAL FEATURES OF NONDISCRETE BIOMOLECULAR DEVICES

Fabrication of biomolecular computing and information processing devices implies preliminary estimation of their design characteristics and

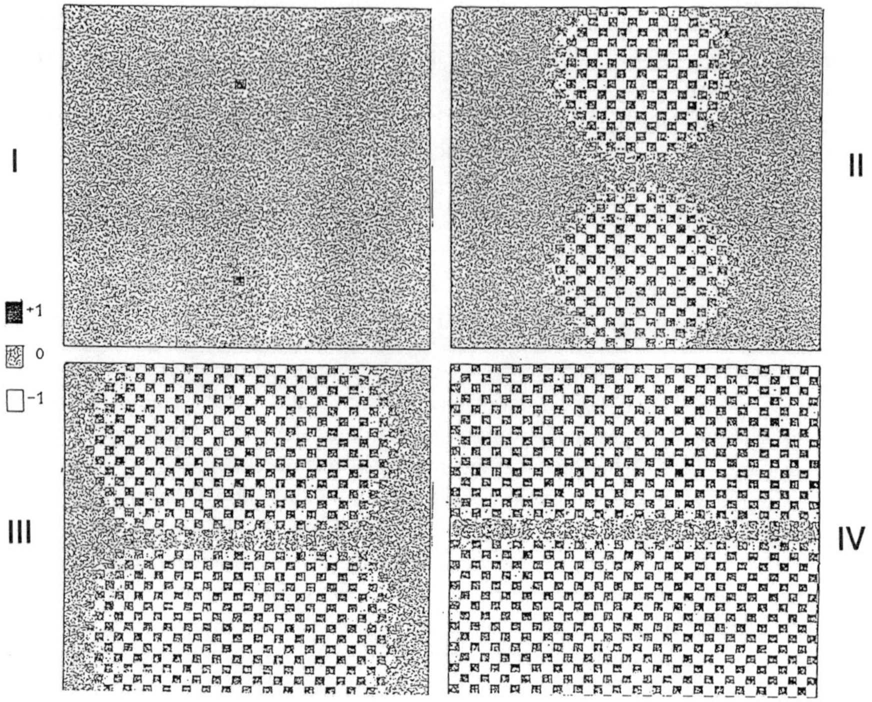

FIGURE 21. Dissipative pattern formation in the distributed multibox dynamic system discussed in the text. I–IV correspond to temporal evolution of the system.

operational features. They are determined by the current state of the theory of biomolecular computing and by the contemporary biomolecular technology. The most essential features are the possible computational load (speed of processing) and possibilities of implementation of proper power supply and continuous functioning. Let us try to discuss these characteristics for the case of biochemical information processing devices.

As it is known, the time needed for elaboration of a measurable amount of product for enzymatic reactions is rather high. As a rule it ranges from some hundredths of a second up to some seconds, being by several orders of magnitude higher than the operational time of the existing semiconductor-based logical elements. The high operational time which characterizes the components constituting the biomolecular dynamic systems as well as the rather low diffusion rates of the molecular species in biochemical media seems to question the expediency of developing nondiscrete biomolecular devices. However, a closer examination of the capabilities and operating features allows one to look at the matter with more enthusiasm.

The use of nondiscrete molecular-based devices intended to solve problems of high computational complexity apparently requires new approaches

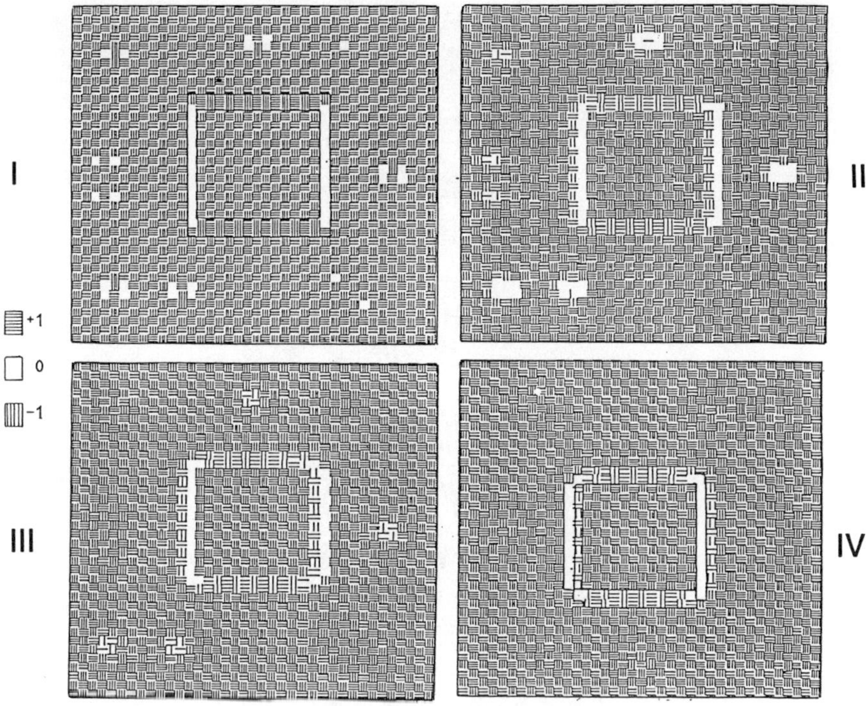

FIGURE 22. Filtering random noises using the distributed multibox dynamic system discussed in the text. I–IV correspond to temporal evolution of the system.

to estimating the speed of processing. Suppose there is a molecular device that is capable of solving some problem intractable by traditional digital computers. In fact in this case, the question is not whether the time within which the solution of the task was obtained is short but is whether this time satisfies the user. In other words, the most important point is the problem of the real time scale inherent in the given problem. For example, the characteristic time scale for economic and sociological problems is several years. Therefore, it does not matter whether the task will be solved within some fractions of a second or even within some days. At the same time, the necessary speed of evaluation of the environmental situation by the pattern recognition system for an autonomously moving vehicle should be within fractions of a second and increase with an increase in vehicle speed.

Thus, in searching for an optimal solution of some problem one should proceed both from the degree of computational complexity of the problem (and possibility of its alternative solution by traditional numerical computers) and from the time scale of problem solution attainable by a molecular device specifically designed for solving this problem.

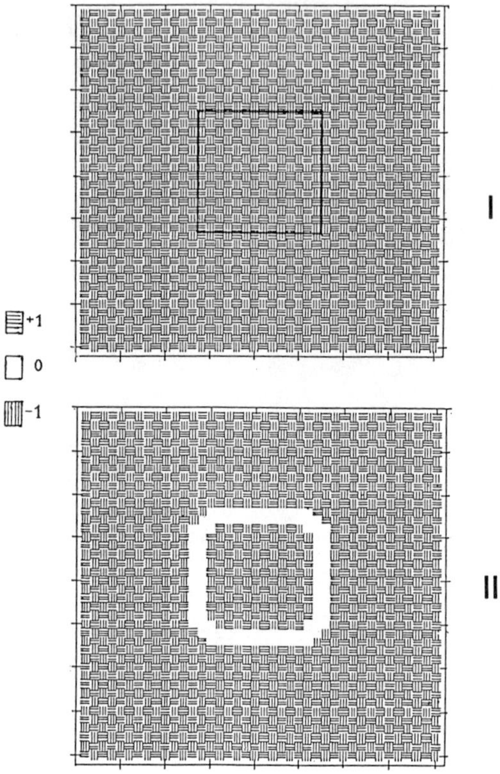

FIGURE 23. Extracting the contour of the square using the distributed multibox dynamic system discussed in the text. I corresponds to the initial state of the system; II corresponds to the final state.

The high parallelism of information processing devices inherent in the devices under consideration allows for high speed of processing. An example is the system representing the planar supramolecular structure on which the photochemically activated enzymes are immobilized. The system can be used for highly parallel summation of two light patterns simultaneously projected onto the surface of the device. The present-day technology allows fabrication of this active media type of about 1 cm^2 in size. Suppose the spatial resolution of the system is about 10 μm which corresponds to the elementary compartment of the pattern of about 10×10 μm. Then the number of compartments on the 1 cm^2 surface will be 10^6. Assuming 1% filling of the surface with enzyme molecules, each compartment will contain 10^4 to 10^5 of them. Let the operational time, i.e., the time of production of a measurable quantity of product be 1.0 to 1.0×10^{-2} s. In this case taking into account that the degree of parallelism is about 10^6, the effective time of a single operation will be of the order of 10^{-6} to 10^{-8} s (using technologically simple primitive computing elements).

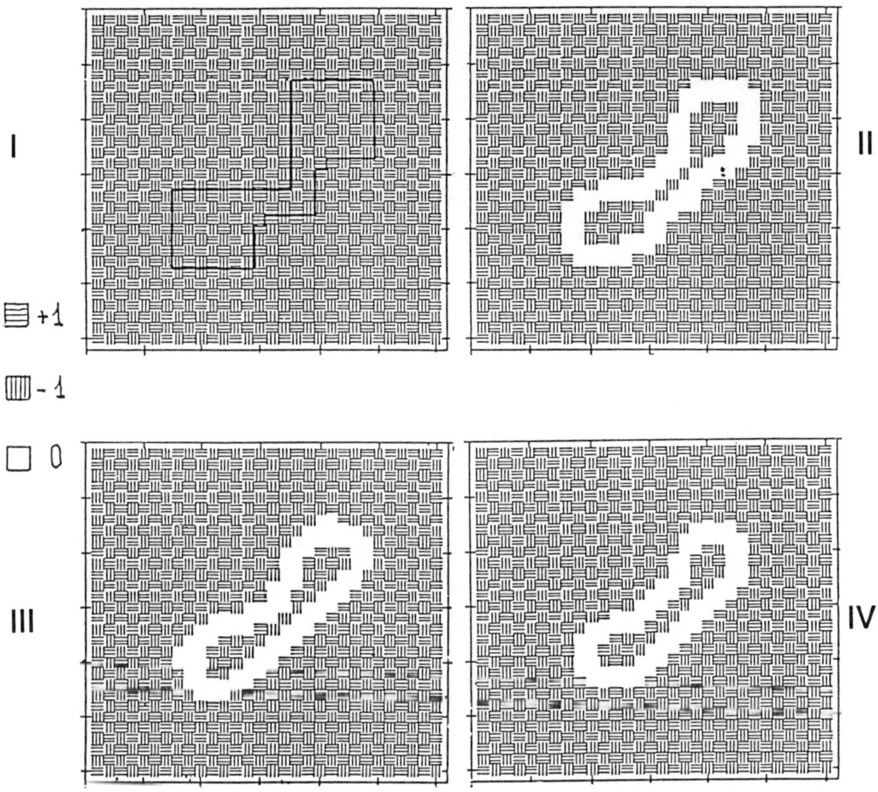

FIGURE 24. Smoothing the contour of the complicated image using the distributed multibox dynamic system discussed in the text. I–IV correspond to temporal evolution of the system.

These operational capabilities comparable with characteristics of contemporary silicon devices are not impressive. However, it is not difficult to see that they correspond to the simplest computational operation, i.e., to the parallel summation of information. Operational capabilities will be much better for computational operations of high complexity, performed by single elementary cells and especially by nonlinear dynamic media as elementary operations (see Rambidi et al.[25] and Rambidi and Chernavskii[26]). This problem will be discussed in more detail in subsequent publications.

The problem of power supply in biomolecular computers should be a subject of special consideration. Since information processing occurs at the molecular level, the required energy should be provided chemically, i.e., by means of a solution of complex composition containing substrates of reactions proceeding at all levels of information processing. The solution is supposed to be pumped through the system at a preset velocity (Figure 16). To provide this the construction should envisage permeability of the system for liquid media. Particularly, the supramolecular system that realizes the level can be formed on a porous surface.

It should be emphasized that pumping of the solution (energy carriers) is important in one more aspect, i.e., it should provide removal of heat produced by energy dissipation in the course of information processing.

The usage of solutions for power supply, though being an uncommon performance characteristic, can hardly be more difficult for implementation in molecular-based devices than when utilizing electricity for power supply in semiconductor-based devices. At the same time, the combination of power supply and heat removing functions seems to be an attractive design feature of biomolecular machines.

Finally, an important point of biomolecular information processing devices is the necessity for restoring the initial state of the dynamic system after each completion of a dynamic process corresponding to a single act of information processing. This problem has not been the subject of investigation up to now. However, based on general considerations the problem does not seem intractable; although from the standpoint of construction, it will be more complicated.

V. THE USAGE OF ACTIVE CHEMICAL MEDIA CHARACTERIZED BY SHORT-RANGE NONLOCAL INTERACTIONS FOR IMAGE PROCESSING

Excitable media characterized by intrinsic nonlocal spatial interactions belong to homogeneous structures with uniformly organized connections between their elements over the whole space. Unlike the classical excitable media, however, they admit interactions not only between the adjacent points, but also in a certain neighborhood of a point. In this section some possible applications of such media are considered for solving problems of image processing discussed in Rambidi et al.,[25] and Rambidi and Chernavskii,[26] and Masterov et al.[77]

A. ACTIVE MEDIA CHARACTERIZED BY NONLOCAL SPATIAL INTERACTIONS

Media having short-range nonlocal interactions between their elements are described by the following system of nonlinear integral-differential equations:[78]

$$\frac{\partial U}{\partial t} = -U + F[-T + \int_D \Phi(\vec{\xi} - \vec{r})U(\vec{\xi},t)d\xi - \beta V]$$

$$\frac{\partial U}{\partial t} = \frac{\tau_U}{\tau_V}(U - V)$$

$$\epsilon = \frac{\tau_U}{\tau_V} \ll 1 \tag{7}$$

where the nonlinear function $F[-T + Z]$ is taken in a stepwise form.

The principal difference between this model and the chemical reaction-diffusion system is that the interaction between elements of the model is described not by a diffusion mechanism, but by a rather long-range function $\Phi(\vec{\xi})$. If $\Phi(\vec{\xi})$ is a rapidly decreasing monotonous function, the system (Equation 7) is practically indistinguishable from ordinary excitable chemical reaction-diffusion media and displays the well-known dynamic properties, such as wave fronts, pulses, etc.[79]

If $\Phi(\vec{\xi})$ is a nonmonotonous function, the equations (Equation 7) generate a class of new rather interesting regimes. For instance, it has generally been recognized up to the present time that one of the basic properties of the waves in classical excitable media is annihilation: that is, two colliding waves cancel each other. In media characterized by nonlocal interactions a different type of dynamics is possible, that is, inhibition of wave propagation at the moment of collision and formation of structures on the line of wave interaction.

Some regimes inherent in the case of the nonmonotonous function $\Phi(\vec{\xi})$ proved to be important in solving problems of image processing. For example, the inhibition of propagation of two colliding waves can evidently be used for skeletonizing images.

Other regimes of image processing using media characterized by the nonmonotonous function $\Phi(\vec{\xi})$ are considered in the next section.

B. POTENTIALITIES FOR IMAGE PROCESSING USING MEDIA CHARACTERIZED BY SHORT-RANGE NONLOCAL INTERACTIONS

A medium having short-range nonlocal interactions described by Equation 7 can be represented by a matrix of processor elements with appropriate spatial connections among them.[77]

It is possible to exemplify some simple logical operations under images performed by such media (Masterov et al.[77]).

1. Transformation of Halftone Images into High-Contrast Ones

This operation can be performed using unconnected processor elements of the trigger type. In this case, however, a ''universal'' matrix of connected processor elements can also be used. To perform threshold processing it is necessary to use nonlocal spatial interactions that will allow stationary of wave fronts. One example of such calculations based on Equation 7 is shown in Figure 25.

2. Contour Enhancement

In the case of the excitable regime of active elements and stationary dynamic spatial structures having a characteristic size approximately equal to the width of the spatial coupling function, the initial stage has more contrast and then its contour is discerned (Figure 26). In this case it is possible to obtain both dark contours against the light background and light contours against the dark background.

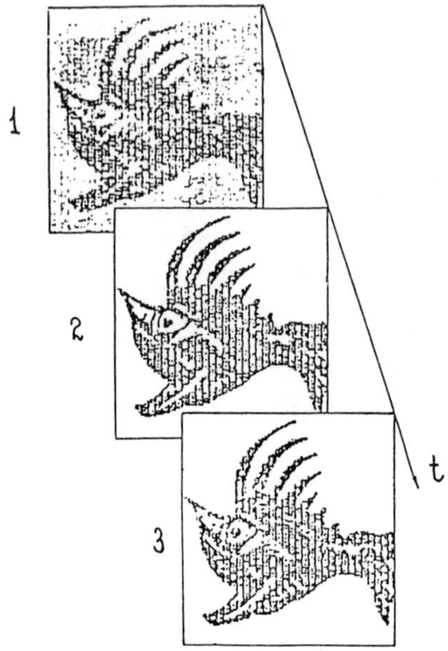

FIGURE 25. The process of changing contrast of an image based on active dynamic media. 1, 2, 3 are steps of temporal evolution.

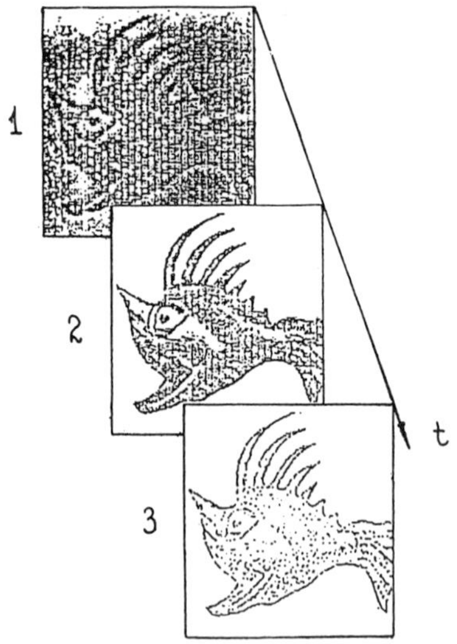

FIGURE 26. The process of contour enhancement based on active dynamic media. 1, 2, 3 are steps of temporal evolution.

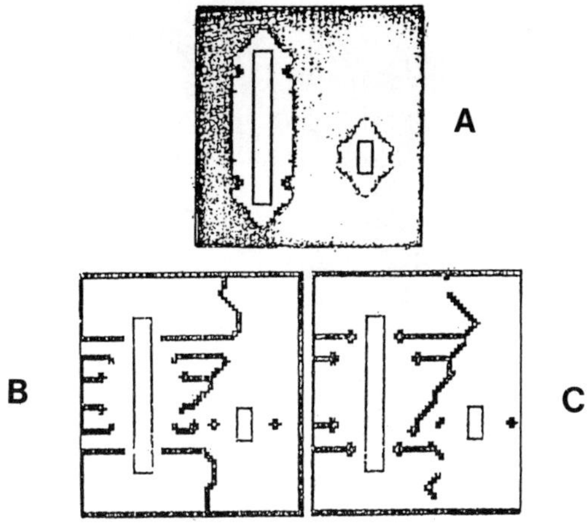

FIGURE 27. The process of image skeletonizing based on active dynamic media (see details in the text).

3. Image Skeletonizing

In a trigger system having a large threshold where the propagating wave fronts reduce the region of the excited state and give rise to stationary dynamic spatial structures in the course of collision, the axial lines of different parts of an image are discerned (the so-called "skeleton" of the image). An illustration of this operation is provided by Figure 27 that shows one of the possible skeletons of the region surrounding two rectangles inscribed into one another.

4. Extraction of Lines Having a Given Direction

If the connection strength in one direction is higher than in the others, only lines of this direction can be extracted during structure generation while the lines of all other directions will disappear. Examples of such extraction of lines (with angles $\varphi = +45°$ and $\varphi = -45°$) are shown in Figure 28.

5. Calculation of Invariant Features

Based on the procedure of extracting lines of given directions it is possible to calculate invariant features of an image. Let $S(\varphi)$ denote the integral intensity of the lines extracted from an image along some direction on the entire plane processed. Let us determine the set of the values $S(\varphi + m\Delta\varphi)$ where the anisotropy angle of the function $\Phi(\vec{\xi},\varphi)$ is changed in steps $\Delta\varphi = \pi/M$ (where M is an integer). Let us consider the function:

$$I\left(m\frac{\pi}{M} + \varphi_0\right) = \frac{S\left(\varphi_0 + m\frac{\pi}{M}\right)}{\sum_{m=0}^{M-1} S\left(\varphi_0 + m\frac{\pi}{M}\right)} \tag{8}$$

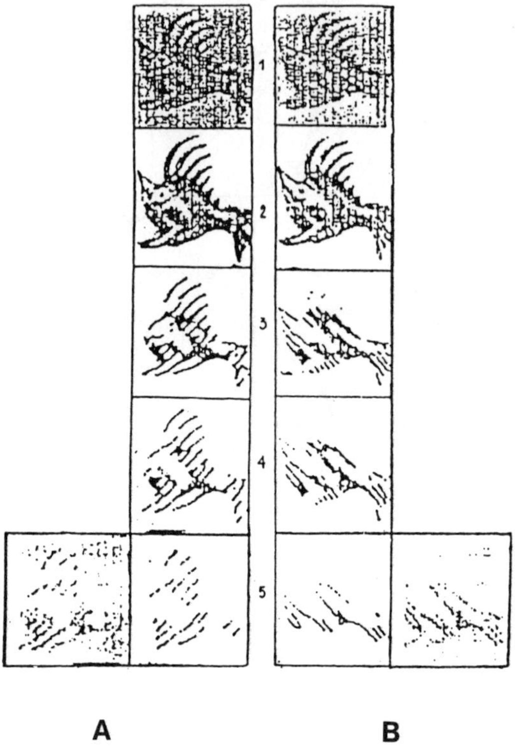

FIGURE 28. Extraction of lines having given directions (see details in the text).

This function is invariant to image location and within a certain range to its size.

In conclusion, it should be noted that further studies of the dynamic structures described by the type of models in Equation 7 may allow better understanding of the results of experiments with biological tissues and formulation of principles for designing architectures for "neuron-type" computers as well as development of efficient methods of programming such highly parallel computers.

C. IMAGE PROCESSING USING LINEAR DIFFUSION

Sibert and Waxman[80] proposed a method for solving some problems of image processing that uses propagation of excitation based on the diffusion mechanism. Consider a pseudoflat region R(X,Y). A contoured halftone image A(R) input into dynamic medium diffuses according to the linear diffusion equation:

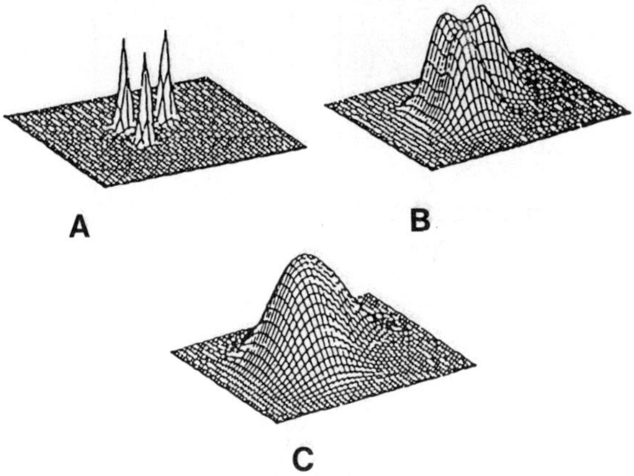

FIGURE 29. Spatial representation of the diffusion process characterized by three point sources. (A) Initial excitations; (B) the process after small temporal interval Δt; (C) the process after a considerable period of time $t >> \Delta T$).

$$\frac{dA}{dt} = æ\,\Delta A(R) - \lambda A(R) + (A_{sat} - A(R)) \sum_i \delta[X - X_i(t), Y - Y_i(t)]$$
$$+ [A_{sat} - A(R)] \sum_m G_{\sigma_1}[X - X_m(t), Y - Y_m(t)]$$
$$- A(R) \sum_m G_{\sigma_2}[X - X_m(t), Y - Y_m(t)] \quad (9)$$

Here æ accounts for the density and conductivity of the region, A_{sat} is the activity saturation value, λ causes a passive decay, δ is the unit impulse function, and G_{σ_1} and G_{σ_2} are normalized Gaussian functions with spreads σ_1 and σ_2.

Subsequent transformation of the maxima of distribution A(R) and of the Gaussian functions permits solving a number of problems as seen in Figures 29–32.

The advantage of the described algorithm is that it can be realized using parallel computers. According to Balkarei et al.[81] VLSI implementation of this algorithm is being designed. This method provides an easy solution for the problem of finding the points of curve intersections and vertices of angles (Figure 32) of halftone images. However, some other problems (for instance, determining the center of gravity of complex geometrical figures) require indefinite, practically infinite, or very long time. Besides, in linear diffusion equations the amplitude drops with time. Keeping its level constant (in fact, renormalization) requires special efforts and is possible only in some selected points of an image.

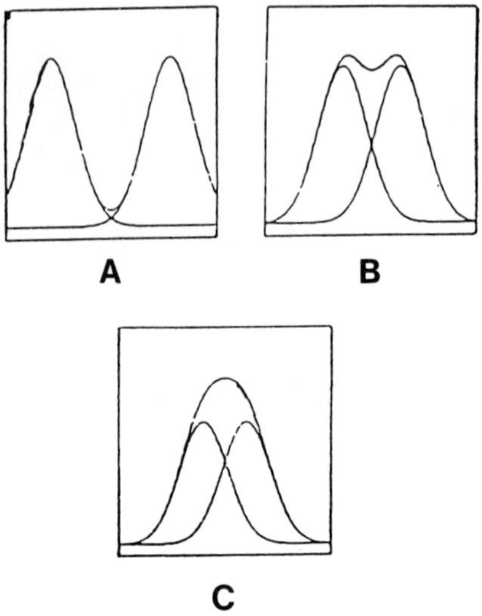

FIGURE 30. Temporal evolution of a diffusion process (see details in the text).

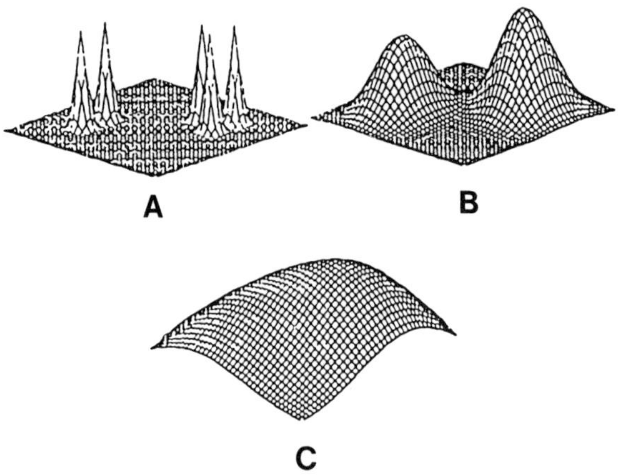

FIGURE 31. Spatial representation of a complicated diffusion process. A, B, C are steps of temporal evolution.

In the next section a method designed to avoid these shortcomings in image processing will be described. An excitable medium is essentially a medium based on diffusion as the media described above. However, it is also based on autocatalytic processes. Because of this difference the waves propagating in such media acquire properties that are basically different from those of classical waves. Their amplitudes do not drop with distance, and they do

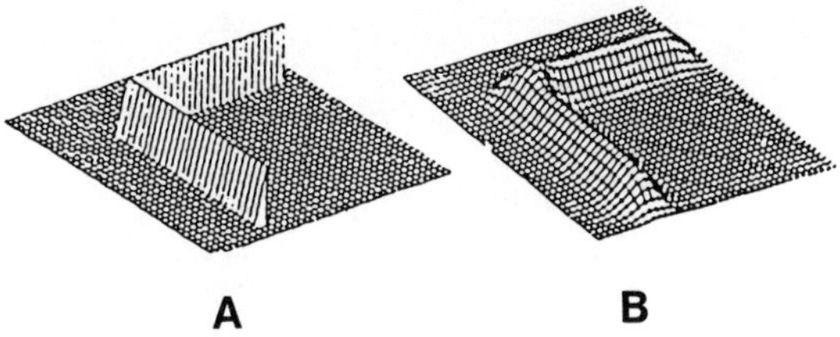

FIGURE 32. Temporal evolution of an angle excitation in the diffusion process. The maximum of distribution coincides with its vertex.

not reverberate from medium boundaries or interfere. Therefore, an excitable medium can be used as a highly parallel processor.

D. LIGHT-SENSITIVE CHEMICAL WAVES AS AN INSTRUMENT FOR IMAGE PROCESSING

Nowadays the speed of processing large information arrays is almost completely dependent on the number of processors in a computer (i.e., on computational parallelism) before the processing speed of a single processor approaches the theoretical limit. However, the cost of computers grows progressively while the number of processors increases. Therefore, for solving a number of problems (including image processing), it would be advantageous to increase the size of the multiprocessor matrix and to use processors as simple as possible.[81] The simplest version of the matrix is a two-dimensional network of locally connected 1-b processors. For such systems a specific class of algorithms is being developed: the so-called wave front algorithms of image processing.[81]

The possibility of implementing image processing methods based on nonlinear dynamic processes now becomes more realistic because natural excitable media are known. The most convenient and best studied among them are the chemical media based on the Belousov-Zhabotinsky reaction.[82] These media consist of a thin (0.2 to 0.7 mm) nonstirred solution layer in which spatiotemporal oscillating reactions occur. Each microvolume of the chemical active medium is a ''chemical trigger'' that can change its color in the course of reaction. The microvolumes are locally coupled by reaction product diffusion. Because of the dynamic properties, the system can display a rich variety of complicated regimes. For image processing using wave processes in chemical active media, it is necessary to have the possibility of inputting (suppose by visible light radiation) an image into the media, i.e., controlling the media parameters at each point of an image by the radiation intensity at the point. The simplest way is to use the light-sensitive modification of the Belousov-Zhabotinsky reaction based on the Ru complex as a catalyst instead of ferroin.[70] In this case it is sufficient to project an image onto the reaction layer surface using properly selected illumination intensity. An experimental setup is shown schematically in Figure 33.

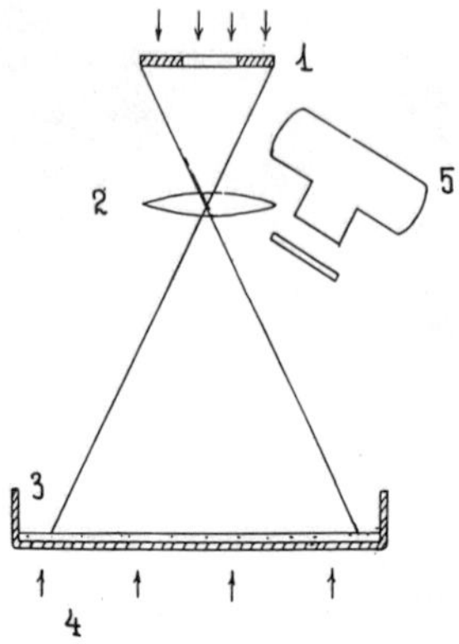

FIGURE 33. The experimental setup for image processing using light-sensitive chemical waves. (1) Initial image; (2) image projection system; (3) active chemical medium; (4) illuminating radiation; (5) camera.

Figure 34 presents time sequences of image processing that demonstrate the beginning and further evolution of an image when its negative picture is projected onto the surface of the Ru-catalyzed active chemical medium which is in the temporal oscillating regime. In frame 1 the image makes its appearance, in frames 2 and 3 its contour enhancement and changes in contrast of the image take place, in frames 4 and 5 the negative to positive transformation of the image occurs, in frame 6 the system comes to the resting state, and in frames 7–12 the cycle recurs.

It is seen that by projecting an image onto a temporally oscillating medium it is possible to obtain a halftone image with a periodically changing contrast. The image passes through the following phases: blank, negative, positive, and blank (the negative image appears first). It has been shown that such a system can be used as a mask to control the contrast.[77]

Figure 35 illustrates contour enhancement for different fragments of an image by active chemical media. Similar to Figure 34 different frames correspond here to different phases of the oscillatory cycle of the system, the only difference being that illumination intensity showed a fourfold increase. In this case the brightest parts of the original negative image are so bright that the projected light not only shifts the phases of oscillations but also causes the transformation from an oscillatory to excitable regime (see Figure 36) by inhibiting the reaction. Therefore, at the boundary of the oscillatory and excitable regions there appears the wave that reveals the contour of the image (frame 4) and propagates from the boundaries to the center of the image (frames 5–12). In frames 7–12 the cycle recurs.

When the projected image has discontinuities (Figure 36), the waves that it initiates collide and merge, thus smoothing the image. If the process lasts a time t, the discontinuities having the length $L_o = {}^1/_2\, vt$ can be smoothed where v is the propagation velocity of waves in the chemical reaction.

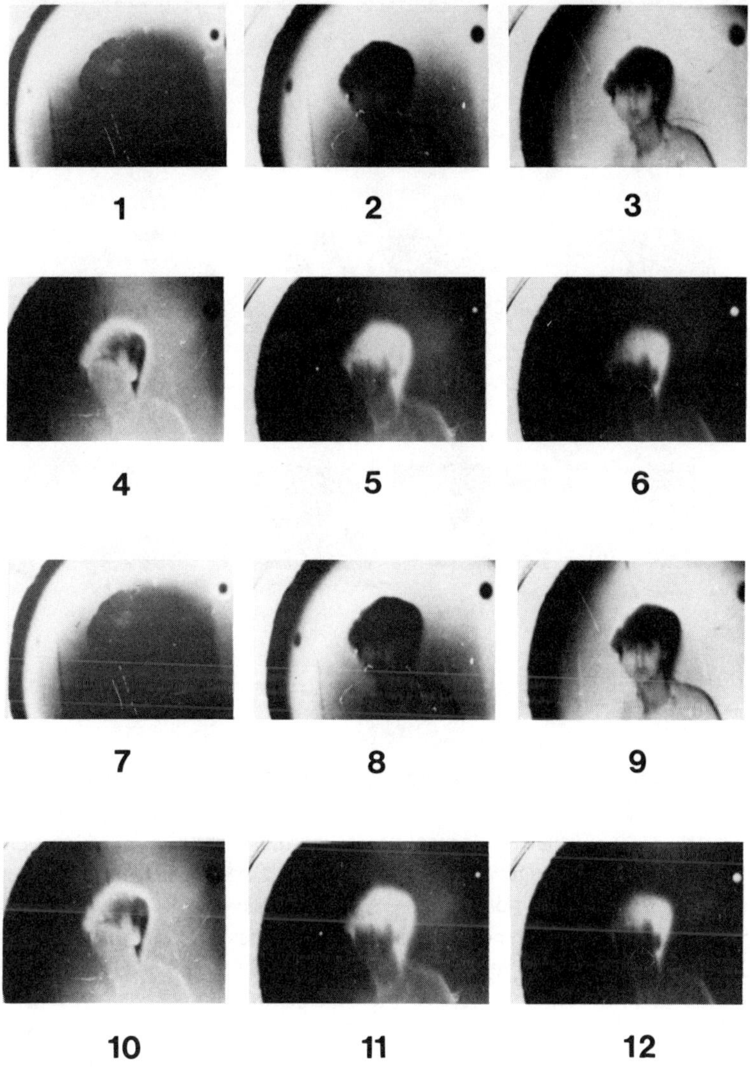

FIGURE 34. Light-sensitive Belousov-Zhabotinsky reaction in the temporal oscillation mode (see details in the text).

Let us consider the simplest mechanisms of image processing by the light-sensitive Belousov-Zhabotinsky reaction. This reaction, as well as the original one proceeds in the oscillatory regime, each point alternating between blue and orange colors with a period of about 1 to 2 min. If a particular arrangement of blue and orange regions is prearranged in the reaction layer (by drawing on the reaction surface with a silver wire), positive to negative transitions

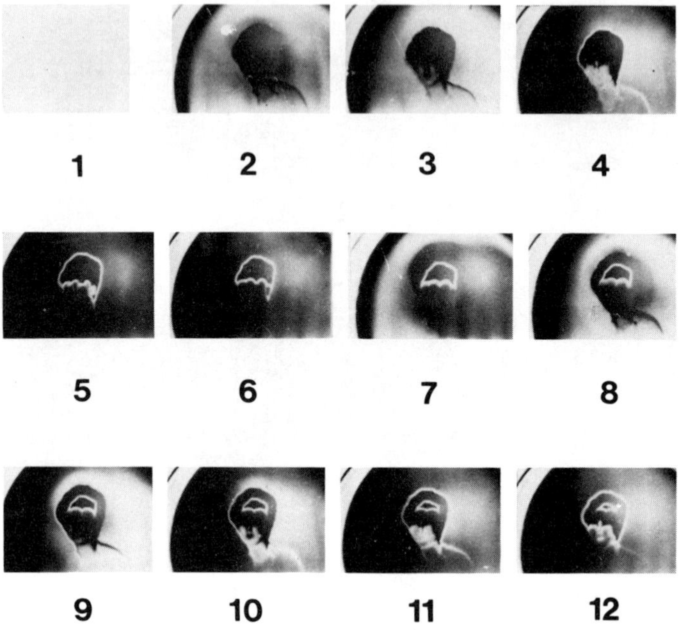

FIGURE 35. Contour enhancement using Belousov-Zhabotinsky reaction (see details in the text).

will occur due to oscillations (the orange color is replaced by blue and vice versa) with a period equal to the reaction oscillation period.

In the light-sensitive BZ reaction the initial picture, instead of being drawn, can be prearranged by light projecting a particular negative image (see Figure 37a, top). Initially when the illumination is switched on no image appears in the reaction layer despite spatial differences in illumination intensity (phase 1). Then since the reaction is light inhibited and the oscillation period increases in the illuminated areas, the change in color occurs first in the nonilluminated areas (phase 2). The initial negative image changes to the positive one and then, because of oscillations, the reverse situation takes place (phase 3) whereupon the process repeats periodically.

In Figure 34 these transformations can be seen for a real image having continuous distribution of illumination intensity. Frame 1 corresponds to termination of phase 1 in Figure 37a, frames 2 and 3 to phase 2, frames 4 and 5 to phase 3, and frame 6 to phase 4. In these transformations the contrast of an image changes smoothly from the maximum value through zero to the negative value (negative image).

By increasing the range of illumination intensity (or by changing the reagent composition), it is possible to see contour enhancement of an image. In the simplest case the contour appears at the boundary between the oscillating

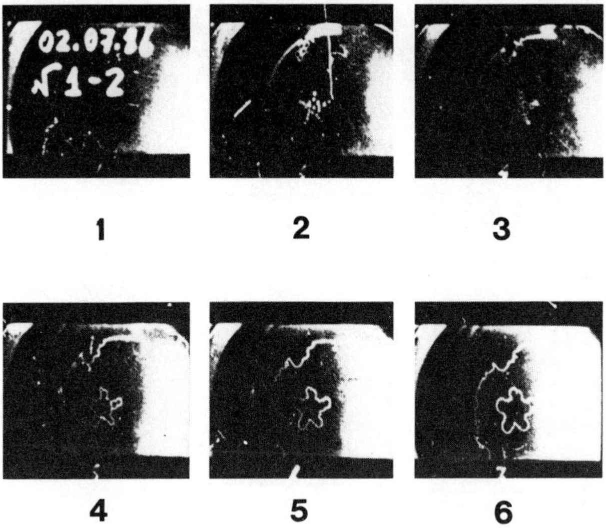

FIGURE 36. Smoothing of a discontinuous image using Belousov-Zhabotinsky reaction (see details in the text).

and excitable regions as illustrated by Figure 37b. The top of the figure schematically shows illumination intensity as a function of the spatial coordinate; below is the resulting distribution of $[Br^-]$.

Br^- is an inhibitor of this reaction and its concentration (c) determines the reaction mode. When $[Br^-] > c_2$, the reaction is completely inhibited; when $c_1 < [Br^-] < c_2$, the medium is excitable; and when $[Br^-] < c_1$, the reaction runs in the oscillatory regime. In the areas where $[Br^-] < c_1$ there are color oscillations with a period equal to the reaction period, and in the areas where $c_1 < [Br^-] < c_2$ the propagating waves appear at the boundary of the oscillating and excitable regions (Figure 37b, bottom).

Figure 35 shows the whole process for the case of an image having continuous distribution of illumination intensity. Note that the chemical waves propagating to the center of the image (frames 9–12) can be used for image skeletonizing.

The propagating waves of the chemical reaction can also be used for smoothing an image. Figure 36 shows different stages of development of the wave pattern resulting from projection of a fragmentary image. The mechanism of smoothing is clear in this case.

The simple examples of image processing presented here were performed by light-sensitive propagating chemical waves. In contrast to many local methods of image processing the propagating waves can also perform nonlocal nonlinear operations in image processing.

It should be noted that Babloyantz and Sepulchre[83] attempted to use the spatial-temporal properties of dynamic systems similar to BZ reaction for

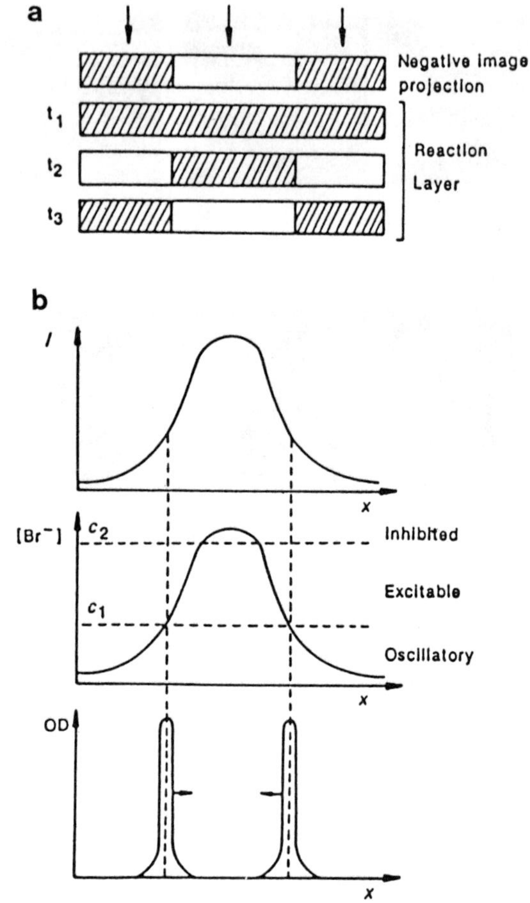

FIGURE 37. Mechanism of image processing. (a) Positive-negative transitions and contrast control; (b) contour enhancement (top — illumination intensity of the projected image [I] and corresponding bromide concentration; bottom — optical density in the reaction layer).

solving artificial intelligence problems. They designed a two-dimensional network of coupled oscillators which can take two different states. When excited the network of oscillators can demonstrate waves resembling those emitted by a source of the leading center type. This property was used by the authors to solve the problem of finding an optimal way in a complicated maze.

These authors also consider networks of excitable elements with nonlinear coupling.

VI. CONCLUSIONS

Does the nondiscrete biomolecular computer have real potential for solving problems of high computational complexity?

Given experimental investigations and results of numerical simulation discussed in this chapter, the answer proved to be more yes than no.

The contemporary level of knowledge is too poor to design and to construct operational models of molecular (biomolecular) devices capable of solving arbitrary complicated problems of practical interest. Nevertheless, potentialities of the approach discussed can be sufficient enough to elaborate operational devices for solving rather simple but practically important tasks of high computational complexity.

Further progress in developing the approach discussed will depend greatly on:

1. Results of future detailed experimental investigations on dynamic biomolecular and biochemical systems having very high behavioral complexity
2. Elaboration of the theoretical approach that will enable us to find "one-to-one" correspondence between behavioral complexity of a dynamic system and computational complexity of problems that can be solved using the system
3. Employing modern preparative methods (based on, for instance, gene engineering thermophylic microorganisms) to produce stable biomacromolecules that possess properties optimal for using them to construct biomolecular devices

REFERENCES

1. **Feynman, R. P.,** There's plenty of room at the bottom, in *Miniaturization,* Gilbert, H. D., Ed., Reinhold, New York, 1961, 25.
2. **Krumhansl, J. A. and Pao, J.-H.,** Microscience: an overview, *Phys. Today,* N11, 25, 1979.
3. **Drexler, K. E.,** Molecular engineering: an approach to the development of general capabilities for molecular manipulation, *Proc. Natl. Acad. Sci. U.S.A.,* 78, 5275, 1981.
4. **Lidorenko, N. S., Ilyn, B. I., and Zaidenman, I. A.,** *Vvedenie v molekulyarnuyu electroniku,* Energoatomizdat, Moscow, 1984 (in Russian).
5. **Carter, F. L., Ed.,** *Molecular Electronic Devices,* Marcel Dekker, New York, 1982.
6. **Carter, F. L., Ed.,** *Molecular Electronic Devices II,* Marcel Dekker, New York, 1987.
7. **Aviram, A., Ed.,** *Molecular Electronics — Science and Technology,* United Engineering Trustees, New York, 1989.

8. **Hong, F. T., Ed.,** *Molecular Electronics: Biosensors and Biocomputers,* Plenum Press, New York, 1989.
9. *Proc. Int. Symp. Future Electron. Devices,* Press Center Hall, Tokyo, 1985.
10. **Rambidi, N. G., Zamalin, V. M., Sandler, Iu. M., Todua, P. A., and Kholmansky, A. S.,** Molecularnaya elementnaya basa perspektivnykh informatsionno-logicheskikh ustroystv, *Itogi nauki i techniki,* VINITI, Moscow, 1987 (in Russian).
11. **Airapetianz, S. V.,** Predelnye vozmozhnosti elektronnykh priborov, in *Biomolekularnaya Elektronika i Problema Samosborki Nadmolekulyarnykh Structur,* Lazarev, P. I., Ed., NZBI AN SSSR, Puschino, 1987 (in Russian).
12. **Rambidi, N. G., Chernavskii, D. S., Krinsky, V. I., Nikolaenko, V. M., Gritsenko, O. V., Sidelnikov, D. I., and Simonova, A. P.,** Nediskretnye biomolekulyarnye informatsionno-logicheskye ustroystva: fizicheskye prinzipy i vozmozhnye puti postroeniya, *Itogi Nauki i Techniki,* VINITI, Moscow, 1991 (in Russian).
13. **Smith, K.,** RISC — microprocessor of Acorn Ltd., *Electronics,* 58, 14, 1985.
14. **Brodie, I. and Muray, J. J.,** *The Physics of Microfabrication,* Plenum Press, New York, 1982.
15. **Gulyaev, Yu. V., Sandomirsky, V. B., Suchanov, A. A., and Tkach, Yu. Ya.,** Fizicheskie Ogranicheniya minimalnykh razmerov elementov sovremennoy mikroelectroniky, *Usp. Fiz. Nauk,* 144, 475, 1984 (in Russian).
16. **Lindbeck, R.,** Firma TI ischet preemnika transistoru, *Elektronika,* 58, 61, 1985 (in Russian).
17. **Shulz, G. E. and Schirmer, R. H.,** *Principles of Protein Structure,* Springer-Verlag, New York, 1979.
18. **Andrew, A. M.,** *Artificial Intelligence,* Abacus Press, Chillaton, Devon, U.K., 1983.
19. **Trachtenbrot, B. A.,** Algoritmy i vychislitelnye avtomaty, *Sovietskoe Radio,* Moscow, 1974 (in Russian).
20. **Conrad, M.,** On design principles for a molecular computer, *Commun. ACM,* 28, 464, 1985.
21. **Conrad, M. and Rosenthal, A.,** Limits on the computing power of biological systems, *Bull. Math. Biol.,* 43, 59, 1981.
22. **Conrad, M.,** Molecular computer design and biological information processing, in *Proc. Int. Symp. Future Dev., Bioelectron. Mol. Electr. Dev.,* Tokyo, 1985.
23. **Rambidi, N. G., Zamalin, V. M., and Sandler, Yu, M.,** Molecular information-processing devices and artificial intelligence problems, *J. Mol. Electron.,* 4, S39, 1988.
24. **Rambidi, N. G.,** On design principles for functionally flexible molecular devices, in *Molecular Electronics — Science and Technology,* Aviram, A., Ed., United Engineering Trustees, New York, 1989, 205.
25. **Rambidi, N. G., Chernavskii, D. S., and Sandler, Yu. M.,** Towards a biomolecular computer. I. Ways, means, objectives, *J. Mol. Electron.,* 7, 105, 1991.
26. **Rambidi, N. G. and Chernavskii, D. S.,** Towards a biomolecular computer. II. Information-processing and computing devices based on biochemical non-linear dynamic systems, *J. Mol. Electron.,* 7, 115, 1991.
27. **Gritsenko, O. V., Sidelnikov, D. I., Simonova, A. P., Rambidi, N. G., and Chernavskii, D. S.,** Towards a biomolecular computer. III. Information-processing features of distributed biochemical systems functioning in the mode of dissipative structures formation, *J. Mol. Electron.,* 7, 155, 1991.
28. **Conrad, M.,** The Price of Programmability, in *The Universal Turing Machine. A Half-Century Survey,* Herken, R., Ed., Verlag Kammerer & Unverzagt, Hamburg-Berlin, 1988, 285.
29. **Conrad, M.,** The brain-machine disanalogy, *BioSystems,* 22, 197, 1989.
30. **Conrad, M.,** Towards the molecular computer factory, in *Molecular Electronics: Biosensors and Biocomputers,* Hong, F. T., Ed., Plenum Press, New York, 1989, 385.

31. **Aviram, A. and Ratner, M. A.,** Molecular rectifiers, *Chem. Phys. Lett.*, 29, 277, 1974.
32. **Schoelter, U., Tews, K. H., and Kuhn, H.,** Potential model of dye molecule from measurements of the photo current in monolayer assemblies, *J. Chem. Phys.*, 61, 5009, 1974.
33. **Kuhn, H.,** Electron transfer in monolayer assemblies, *Pure Appl. Chem.*, 51, 341, 1979.
34. **Panetta, Ch. A., Baghdadchi, J., and Metzger, R. M.,** TTF-$NHCO_2(CH_2)_2O$-TCNQBr and TTF-$CO_2(CH_2)_2O$-TCNQBr, two potential molecular rectifiers, *Mol. Cryst. Liq. Cryst.*, 107, 103, 1984.
35. **Metzger, R. M. and Panetta, Ch. A.,** Toward organic rectifiers, in *Molecular Electronic Devices II*, Carter, F. L., Ed., Marcel Dekker, New York, 1987, 5.
36. **Metzger, R. M. and Panetta, Ch. A.,** Langmuir-Blodgett films of potential organic rectifiers, in *Molecular Electronics — Science and Technology*, Aviram, A., Ed., United Engineering Trustees, New York, 1989, 293.
37. **Aviram, A., Seiden, P. E., and Ratner, M. A.,** Theoretical and experimental studies of hemiquinones and comments on their suitability for molecular information storage elements, in *Molecular Electronic Devices*, Carter, F. L., Ed., Marcel Dekker, New York, 1982, 5.
38. **Aviram, A.,** Preparation of monomolecular layers of hemiquinones for electric field switching, *J. Mol. Electron.*, 4, S99, 1988.
39. **Aviram, A., Joachim, C., and Pomerantz, M.,** Evidence of switching and rectification by a single molecule affected with a scanning tunneling microscope, *Chem. Phys. Lett.*, 146, 490, 1988.
40. **Carter, F. L.,** Conformational switching at the molecular level, in *Molecular Electronic Devices*, Carter, F. L., Ed., Marcel Dekker, New York, 1982, 51.
41. **Carter, F. L., Schultz, A., and Duckworth, D.,** Soliton switching and its implications for molecular electronics, in *Molecular Electronic Devices II*, Carter, F. L., Ed., Marcel Dekker, New York, 1987, 149.
42. **Ricco, S. and Azbel, M. Ya.,** Tunneling through a multiwell one dimensional structure, *Phys. Rev.*, 29, 4356, 1984.
43. **Pschenichnov, E. A.,** O tunnelnom prochojdenii cheres sistemu odinakovykh potenzialnykh barierov, *Fiz. Tverd. Tela*, 4, 113, 1963 (in Russian).
44. **Tsu, R. and Esaki, L.,** Tunneling in a finite superlattice, *Appl. Phys. Lett.*, 22, S62, 1973.
45. **Shewchuk, T. J., Chaplin, P. C., and Coleman, P. D.,** Resonant tunneling oscillations in a $GaAs_xGa_{1-x}As$ heterostructure at room temperature, *Appl. Phys. Lett.*, 46, 508, 1985.
46. **Carter, F. L.,** Molecular level fabrication techniques and molecular electronic devices, *J. Vac. Sci. Technol.*, 1, 959, 1983.
47. **Davydov, A. S.,** *Solitony v molekulyarnykh sistemakh*, Naukova Dumka, Kiev, 1986 (in Russian).
48. **Davydov, A. S.,** *Solitony v bioenergetike*, Naukova Dumka, Kiev, 1986 (in Russian).
49. **Brazovskii, S. A. and Kirova, N. A.,** Electron self-location and periodic superstructures in quasi-one-dimensional dielectrics, *Soviet Sci. Rev.*, N5, 99, 1984.
50. **Pople, S. A. and Walsmley, S. H.,** Bond alteration defects in long polyene molecules, *Mol. Phys.*, 5, 15, 1962.
51. **Su, W. F., Schrieffer, J. R., and Heeger, A. G.,** Soliton excitation in polyacetylene, *Phys. Rev., Sect. B*, 22, 2099, 1980.
52. **Groves, M. P.,** Dynamic circuits diagrams for some soliton switching devices, in *Molecular Electronic Devices II*, Carter, F. L., Ed., Marcel Dekker, New York, 1987, 183.
53. **Roth, S.,** Polyene chains in molecular electronics, *Z. Phys. Chem., Abt. B*, 64, 25, 1986.
54. **Vanderbilt, D. and Mele, E. J.,** Effects of disorder on the electronic structure of undoped polyacetylene, *Phys. Rev., Sect. B*, 22, 3939, 1986.

55. **Aviram, A.,** Molecules for memory, logic and amplification, *J. Am. Chem. Soc.*, 110, 5687, 1988.
56. **Moerner, W. E.,** Molecular electronics for frequency domain optical storage: persistent spectral hole burning — a review, *J. Mol. Electron.*, 1, 55, 1985.
57. **Birge, R. R., Zhang, Ch.-F., and Lawrence, A. F.,** Optical random access memory based on bacteriorhodopsin, in *Molecular Electronics — Biosensors and Biocomputers,* Hong, F. T., Ed., Plenum Press, New York, 1989, 369.
58. **Wild, U. P. and Renn, A.,** Molecular computing: a review. I. Data and image storage, *J. Mol. Electron.*, 7, 1, 1991.
59. **Nicolis, J. S.,** *Dynamics of Hierarchical Systems. An Evolutionary Approach,* Springer-Verlag, Berlin, 1986.
60. **Chernavskii, D. S.,** *Synergetika i Informatika,* Znanie, Moscow, 1990 (in Russian).
61. **Rambidi, N. G.,** Obschie printsipy pererabotky informatsii v slozhnykh molekulyarnykh sistemakh, in *Itogi Nauki i Techniki,* VINITI, Moscow, 1990 (in Russian).
62. **Markin, V. S., Pastushenko, V. F., and Chizmadzhev, Yu. A.,** *Teoriya Vozbudimykh Sred,* Nauka, Moscow, 1989 (in Russian).
63. **Krinsky, V. I., Medvinsky, A. B., and Panfilov, A. V.,** *Evolutsiya Avtovolnovykh Vikhrey,* Znanie, Moscow, 1986 (in Russian).
64. **Romanovsky, Yu. M., Stepanova, N. V., and Chernavskii, D. S.,** *Matematicheskaya Biofizika,* Nauka, Moscow, 1984 (in Russian).
65. **Gaididey, Yu. B., Harkianen, V. N., and Chinarov, V. A.,** Synergeticheskaya model funkzionirovaniya potenzialo-zavisimykh ionnykh kanalov v biomembranakh, Prepr. ITF 88-123P, 1988 (in Russian).
66. **Nicolis, G. and Prigogine I.,** *Self-Organization in Nonequilibrium Systems. From Dissipative Structures to Order through Fluctuations,* John Wiley & Sons, New York, 1977.
67. **Pollak, L. S. and Michailov, A. S.,** *Samoorganizatsiya v Nerav–Novesnykh Fiziko-Khimicheskikh Sistemakh,* Nauka, Moscow, 1983 (in Russian).
68. **Emeljanov, S. V., Rambidi, N. G., and Maximychev, A. V.,** *Tendentsii i Puti Razvitiya Sredstv Pererabotki Informatsii na Molekulyarnom Urovne,* IRIMS, Moscow, 1991 (in Russian).
69. **Balkarei, Yu. I., Evtichov, M. G., and Elinson, M. I.,** Avtovolnovye sredy — raspredelennye analogi neironnykh setey, Prepr. FIAN N1 (530), Moscow, 1990 (in Russian).
70. **Kuhnert, L., Agladze, K. I., and Krinsky, V. I.,** Image processing using light-sensitive chemical waves, *Nature (London),* 337, 244, 1989.
71. **Sandler, Yu. M.,** Model of neural nets with selective memorization chaotic behavior, *Phys. Lett. A,* 144, 462, 1990.
72. **Sandler, Yu. M. and Artyshkin, V. F.,** The neural networks with selective learning and quasi-chaotic attractors, *J. Nonl. Biol.*, 1, 3, 1990.
73. **Kohonen, T.,** *Self-Organization and Associative Memory,* Springer-Verlag, Berlin, 1984.
74. **Grossberg, S.,** Nonlinear neural networks: principles, mechanics and architecture, *Neural Netw.*, 1, 17, 1988.
75. **Sidelnikov, D. I., Gritsenko, O. V., Simonova, A. P., Rambidi, N. G., Polezhaev, A. A., and Chernavskii, D. S.,** Nonlinear dynamics of the distributed biochemical systems functioning in the dissipative structure formation mode, *Biol. Cybern.*, in press.
76. **Naparstek, A., Romette, J. L., Kernevez, J. P., and Thomas, D.,** Memory in enzyme membranes, *Nature (London),* 249, 490, 1974.
77. **Masterov, A. V., Rabinovich, M. I., Tolkov, V. N., and Yachno, V. G.,** Issledovaniye rezhimov vzaimodeistviya avtovoln i avtostructur v neironopodobnykh sredakh, in *Kollektivnaya Dinamika Vozbuzhdeniy i Strukturo-Obrazovaniye v Biologicheskikh Tkanyakh,* IPF AN SSSR, Gorykiy, 1989 (in Russian).
78. **Golubev, V. N., Rabinovich, M. I., Tolkov, V. N., and Yachno, V. G.,** Dinamicheskiye struktury v dvuch-komponentnykh vozbudimykh sredakh. Modelirovaniye na kletochnom komputere, in *Biomolekulyarnaya Elektronika,* NCBI AN SSSR, Puschino, 1988 (in Russian).

79. **Vasilyev, V. A., Romanovsky, Yu. M., and Yachno, V. G.,** *Avtovolnovye Prozessy,* Nauka, Moscow, 1987 (in Russian).
80. **Seibert, M. and Waxman, A. M.,** Spreading activation layers, visual saccades and invariant representation for neural pattern recognition systems, *Neural Netw.*, 2, 9, 1989.
81. **Balkarei, Yu. I., Golik, L. L., and Elinson, M. I.,** *Avtovolnovye sredy,* Znanie, Moscow, 1985 (in Russian).
82. **Zaikin, A. N. and Zhabotinsky, A. M.,** Concentration wave propagation in two-dimensional liquid-phase self-oscillating systems, *Nature (London),* 225, 535, 1970.
83. **Babloyantz, A. and Sepulchre, J. A.,** A network of oscillators can perform ''intelligent'' tasks, in *Nonlinear Wave Processes in Excitable Media,* Holden, A. V., Markus, M., and Othmer, H. G., Eds., Dortmund, 1989.

Chapter 4

MOLECULAR NONLINEAR OPTICAL SWITCHING

Shammai Speiser

TABLE OF CONTENTS

0-8493-8061-8/93/$0.00 + $.50

I. INTRODUCTION

Since the advent of lasers in 1960, nonlinear optics has emerged as an important field of research. Nonlinear optical phenomena and devices based on implementation of them are an integrated part of laser physics research.[1] Progress in optical information processing and computing[2] depends to a large extent on the availability of fast and sensitive molecular optical switches such as spatial light modulators (SLMs).[3] In recent years this field has advanced tremendously; however, the lack of appropriate materials to implement theoretical design of optical hardware is often considered a bottleneck in realizing the potential of the field.[4] Organic materials may provide the ideal medium needed for this optical hardware.[5] It is already known that such materials exhibit high nonlinear optical susceptibilities that are necessary for optical data processing.

So far, most experiments on organic nonlinear optical materials have concentrated on second harmonic generation (SHG);[5,6] however, in recent years optical modulation,[7] optical bistability,[7] optical phase conjugation,[8] and related phenomena have been studied in these systems.

An all-optical device based on appropriate organic materials can operate on the intensity-induced changes in the complex index of refraction, i.e., the absorptive and dispersive properties of the medium due to intensity-induced population changes. It has been shown that optical bistability can be observed in such materials.[9] In a preliminary study, we have examined the possibility of spatial light modulation by using the same physical properties of organic materials which are instrumental in optical bistable response.[10] A steady-state kinetic analysis, ignoring propagation effects, was first used to evaluate average SLM properties of a particular molecular nonlinear absorber. The calculated SLM performance indicates that a molecular SLM can be constructed. More recently we presented a complete theoretical calculation, which includes propagation effects, for the operation and performance of nonlinearly absorbing molecular SLMs.[11] Several switching schemes were then examined experimentally to test the prediction of this calculation.[12] Recently, the theoretical analysis was extended to include time-dependent phenomena.[13]

In the present review chapter we summarize progress made in recent years in this new emerging discipline.

II. MOLECULAR NONLINEAR OPTICAL RESPONSE

A. GENERAL FORMULATION

The nonlinear optical response of a molecular system can be traced to the electrical field intensity ($\vec{E}$) dependence of the molecular dipole moment $\vec{\mu}$:

$$\vec{\mu} = \vec{\mu}(\vec{E}) \tag{1}$$

Since the exact functional form of $\vec{\mu}(\vec{E})$ cannot be calculated for most systems, the alternative approach of expanding $\vec{\mu}(\vec{E})$ in a power series of E is widely used:[1]

$$\vec{\mu} = \vec{\mu}_0 + \alpha \cdot \vec{E} + \beta : \vec{E}\vec{E} + \gamma : \vec{E}\vec{E}\vec{E} + \ldots \tag{2}$$

Equation 2 expresses the fact that application of a static electrical field to a molecule will result in a displacement of the electric charges, thus modifying the molecular dipole moment. In the expansion in Equation 2 the coefficients are the molecular polarizability tensors — α, β, and γ of the first, second, or third order, respectively — the latter two sometimes being called hyperpolarizabilities. Their magnitudes are in the range $\alpha \sim 10^{-23}$ cm^3, $\beta \sim 10^{-30}$ ESU, and $\gamma \sim 10^{-37}$ ESU.[5] For our discussion we should consider frequency-dependent polarizabilities so that Equation 2 should be extended to include time-dependent electrical fields such as those carried by laser beams. For monochromatic fields:

$$\vec{\mu} = \alpha(\omega) \cdot \vec{E}(\omega) + \beta(-\omega;\omega,\omega_2) : \vec{E}(\omega_1)\vec{E}(\omega_2) + \gamma(-\omega;\omega_1',\omega_2',\omega_3') : \vec{E}(\omega_1')\vec{E}(\omega_2')\vec{E}(\omega_3') \tag{3}$$

where the frequencies satisfy the condition:

$$\omega = \omega_1 + \omega_2 = \omega_1' + \omega_2' + \omega_3' \tag{4}$$

thus the second β term in Equation 3 describes, in general, a three-wave interaction while the third γ term generally describes a four-wave interaction.

In general, the molecular polarizability tensors include many components the number of which can be reduced according to the particular symmetry of the molecular system. Perturbation theory can be used to calculate molecular polarizabilities; however, the success of such calculations relies on the availability of exact wave functions. Examples of such calculations are available.

For macroscopic systems Equation 2 has to be averaged in order to obtain the macroscopic counterparts of the microscopic optical linearities. For real systems we need to evaluate the macroscopic polarization $\vec{P}$ which is the average dipole moment per unit volume given by:[1]

$$\vec{P} = \vec{P}_0 + \chi^{(1)} \cdot \vec{E} + \chi^{(2)} : \vec{E}\vec{E} + \chi^{(3)} : \vec{E}\vec{E}\vec{E} \tag{5}$$

where $X^{(n)}$ is the electrical susceptibility tensor of order n which is related to the polarizability of order n, $\alpha^{(n)}$, through:

$$\chi^{(n)}(\omega) = N\alpha^{(n)}(\omega) \prod_{i=1}^{n} f_n \tag{6}$$

TABLE 1
Examples of Electric Susceptibility Tensors Component of $\chi^{(2)}$, $\chi^{(3)}$ for Various Types of Interacting Field Components Together with Associated Phenomena and Known Applications

Susceptibility	Phenomena	Application
$\chi^{(2)}(0; \omega, -\omega)$	Optical rectification	Hybrid bistable device[25]
$\chi^{(2)}(-\omega; \omega, 0)$	Electrooptic (Pockels)	Modulators, variable[1] phase retarders
$\chi^{(2)}(-2\omega; \omega, \omega)$	Frequency doubling	Harmonic generation[1]
$\chi^{(3)}(-\omega_0; \omega_a, \omega_b, -\omega_b)$	AC electrooptic effect, Brillouin scattering, Raman scattering	High-speed optical gates[1]
$\chi^{(3)}(-\omega; \omega, \omega, -\omega)$	AC electrooptic effect AC Kerr effect Self-focusing Degenerate four-wave mixing	Optical bistability,[25] phase conjugation,[47] image processing,[2] real time holography,[57] optical transistors[31]

where N is the number density of molecules and f_n are the local field correction factors which generally are frequency dependent. For isotropic liquids or for cubic crystals the expressions for $\chi^{(1)}$, $\chi^{(2)}$, and $\chi^{(3)}$ read:

$$\chi^{(1)}(\omega) = N\alpha(\omega)f(\omega) \tag{7}$$

where

$$f(\omega) = [\epsilon(\omega) + 2]/3 \tag{8}$$

the well-known Lorentz correction factor for media whose dielectric constant is $\epsilon(\omega)$:

$$\chi^{(2)}(-\omega;\omega_1,\omega_2) = f(\omega)f(\omega_1)f(\omega_2)N\beta(-\omega;\omega_1,\omega_2) \tag{9}$$

$$\chi^{(3)}(-\omega;\omega_1',\omega_2',\omega_3') = f(\omega)f(\omega_1')f(\omega_2')f(\omega_3')N\gamma(-\omega;\omega_1',\omega_2',\omega_3') \tag{10}$$

Table 1 summarizes the various optical phenomena associated with the nonlinear susceptibilities.

Another widely used nonlinear optical property is the nonlinear index of refraction defined by:[1]

$$\eta_{NL} = \eta_0 + \eta_2 I \tag{11}$$

where η_0 is the intensity independent index of refraction and I is the light intensity given (in MKS units) by:

$$I = \frac{1}{2} \eta_0 \epsilon_0 C \langle \vec{E} \cdot \vec{E} \rangle \tag{12}$$

From this relation we can obtain the relationship between η_2 and $\chi^{(3)}$:

$$\eta_2(\text{cm}^2/\text{KW}) = \frac{1}{3}\left(\frac{4\pi}{\eta_0}\right)^2 \chi^{(3)}(\text{esu}) \tag{13}$$

or

$$\eta_2(\text{MKS}) = \chi^{(3)}(\text{MKS})/\eta_0\epsilon_0 \tag{14}$$

In general, η_{NL} is a complex quantity and should be expressed by:

$$\eta_{NL} = \eta_{NL}^{D}(I) + i\eta_{NL}^{A}(I) \tag{15}$$

where the superscripts D and A refer to dispersive and absorptive contributions to η_{NL}, respectively.

B. NONLINEAR COMPLEX INDEX OF REFRACTION FOR A LARGE MOLECULE

Next we consider a nonlinear medium consisting of an ensemble of large organic molecules. This type of molecule has a promising potential for constructing nonlinear optical materials due to the possibility of obtaining desired parameters through organic synthesis. The variety of excitation routes in these molecules give rise to specific intensity dependence of the complex nonlinear index of refraction.

We consider here a molecular absorber where both singlet S and triplet T electronic-state manifolds are involved. The relevant levels and rate parameters are depicted in Figure 1a. The dynamics of this system can be described using the rate-equation approximation:

$$\partial\vec{N}/\partial t = \hat{O} \cdot \vec{N} \tag{16}$$

where $\vec{\mathbf{N}}$ is the population vector given by transposed form:

$$\tilde{N} = (N(S_0), N(S_1), N(S_n), N(T_1), N(T_n)) \tag{17}$$

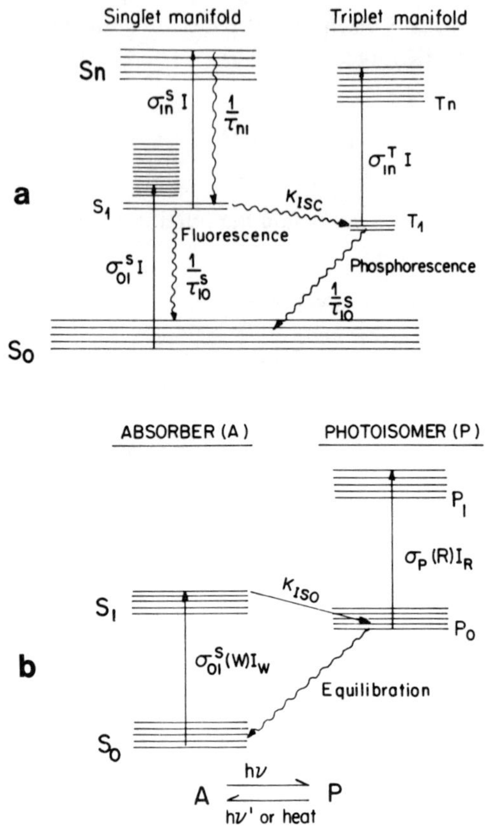

FIGURE 1. (a) Schematic energy level diagram of a nonlinear molecular absorber; (b) schematic energy level diagram of a nonlinear photochromic system absorber.

and the rate constants operator Ô is given by:

$$\hat{O} = \begin{bmatrix} -\sigma_{01}^{S}I & 1/\tau_{10}^{S} & 0 & 1/\tau_{10}^{T} & 0 \\ \sigma_{01}^{S}I & -[1/\tau_{10}^{S} + k_{ISC} + \sigma_{1n}^{S}I] & 1/\tau_{n1}^{S} & 0 & 0 \\ 0 & \sigma_{1n}^{S}I & -1/\tau_{n1}^{S} & 0 & 0 \\ 0 & k_{ISC} & 0 & -(\sigma_{1n}^{T}I + 1/\tau_{10}^{T}) & 1/\tau_{n1}^{T} \\ 0 & 0 & 0 & \sigma_{1n}^{T}I & -1/\tau_{n1}^{T} \end{bmatrix} \tag{18}$$

with $\tau_{ij}^{S,T}$ the decay time of the i level to the j level in the S or T manifolds, respectively; k_{ISC} the intersystem crossing rate constant, $\sigma_{ij}^{S,T}$ the absorption cross section for the i → j transition in the S or T manifolds, respectively;

and I the local light intensity. All states excited higher than the lower excited state are denoted by n; the ground state is S_0.

For continuous wave (cw) or even for nanosecond-pulse excitation, a steady-state solution, $\partial\vec{N}/\partial t = 0$, of Equation 15 is sufficient,[11] yielding the intensity-dependent populations:

$$\vec{N}(I) = [N(S_0)/(k_{ISC} + 1/\tau^S_{10})] \begin{bmatrix} k_{ISC} + 1/\tau^S_{10} \\ \sigma_{01}\tau^S_{10}I \\ \sigma^S_{01}\tau^S_{10}\sigma^S_{1n}\tau^S_{n1}I^2 \\ \sigma^S_{01}k_{ISC}\tau^T_{10}I \\ \sigma^S_{01}k_{ISC}\tau^T_{10}\sigma^T_{1n}\tau^T_{n1}I^2 \end{bmatrix} \tag{19}$$

where the population of S_0 is given by

$$\begin{aligned} N(S_0) = N/[1 &+ \sigma^S_{01}\tau^S_{10}(1 + k_{ICS}\tau^T_{10})I/(1 + k_{ISC}\tau^S_{10}) \\ &+ \sigma^S_{01}\tau^S_{10}(\sigma^S_{1n}\tau^S_{n1} + k_{ISC}\tau^T_{10}\sigma^T_{1n}\tau^T_{n1})I^2(1 + k_{ISC}\tau^S_{10})] \end{aligned} \tag{20}$$

and N is the total molecular concentration.

The origin of the optical nonlinearity in these molecules is the intensity-dependent population. Each level is characterized by an absorption cross section and by an index of refraction which are functions of the excitation wavelength, i.e., their specific electronic spectra. The change in these parameters between different levels is due mainly to contribution of specific resonances from the particular level. The nonlinear complex index of refraction for such a molecular absorber can be written as the scalar product:

$$\eta(I) = \vec{N}(I) \cdot \vec{\eta}' + i(\lambda_0/4\pi)\vec{N}(I) \cdot \vec{\sigma} \tag{21}$$

where the molecular contribution for the real part of the refractive index (dispersive contribution) $\vec{\eta}' = \vec{\eta}^D/N$ is given by its transpose form:

$$\tilde{\eta}' = (\eta'_{S_0}, \eta'_{S_1}, \eta'_{S_n}, \eta'_{T_1}, \eta'_{T_n}) \tag{22}$$

The absorption cross sections [absorptive contribution to $\vec{\eta}(I)$] are given by:

$$\tilde{\sigma} = (\sigma^S_{01}, \sigma^S_{1n}, 0, \sigma^T_{1n}, 0) \tag{23}$$

assuming negligible population for levels $n > 1$. Using Equations 19 and 23, the absorption coefficient can be obtained from Equation 21:

$$\vec{N}(I) \cdot \vec{\sigma} = \alpha(I) = \alpha_0(1 + BI)/(1 + CI + DI^2) \tag{24}$$

TABLE 2
Definition of Rate Parameters Utilized in the Kinetic Analysis of the Nonlinear Molecular Absorber

α_0^W	$\equiv$	$N\sigma_{01}^s(W)$
B	$\equiv$	$\tau_{10}^S[\sigma_{1n}(W) + k_{ISC}\tau_{10}^T\sigma_{1n}(W)]/K$
C	$\equiv$	$\sigma_{01}^S(W)\tau_{10}^S(1 + k_{ISC}\tau_{10}^T)/K$
D	$\equiv$	$\sigma_{01}^S(W)[\sigma_{1n}^S(W) + k_{ISC}\tau_{10}^T\tau_{1n}^T(W)\tau_{n1}^T]/K$
α_0^R	$\equiv$	$N\sigma_{01}^S(R)$
β	$\equiv$	$\xi\tau_{10}^S[\sigma_{1n}^S(R) + k_{ISC}\tau_{10}^T\sigma_{1n}^T(R)]/K$
β'	$\equiv$	$\tau_{10}^S[\sigma_{1n}^S(R) + k_{ISC}\tau_{10}^T\sigma_{1n}^T(R)]/K$

where the various coefficients in Equation 24 are defined in Table 2 for various excitation routes.

The real part of the refractive index is obtained as a weighted sum of the indices of S_0, S_1, and T_1 assuming that S_n and T_n are not significantly populated. Thus the real part of $\eta(I)$ is

$$\eta^D = (A' + B'I)/(1 + CI + DI^2) \tag{25}$$

where A′ and B′ are given in Table 2 for various excitation routes.

III. PROPAGATION EFFECTS AND NONLINEAR OPTICAL SWITCHING

A. SYSTEMS WITHOUT OPTICAL FEEDBACK: MOLECULAR SPATIAL LIGHT MODULATORS

1. Write (Pump) Beam Propagation

In order to analyze propagation effects for a write beam (of intensity I_W), which also serves as the pumping source for the molecular absorber of Figure 1a, we need to solve the dynamics of this system for a write beam propagation in the x direction. This can be described by using the coupled rate equation approximation:

$$\frac{\partial \vec{N}(I_W,x,t)}{\partial t} = \hat{\vec{0}} \cdot \vec{N}(I_W,x,t) \tag{26a}$$

$$\frac{\partial I_w}{\partial x} = -\alpha_W(I_W,x,t)I_W \tag{26b}$$

The solution of Equation 26 yields:

$$\alpha(I_W,x) = \alpha_0^W(1 + BI_W)/(CI_W1 + C_W + DI_W^2) \tag{27}$$

Information is written in the absorbing medium by the absorption of the propagating write beam. Using Equation 27 we can rewrite Equation 26b as:

$$dI_W(1 + CI_W + DI_W^2)/I_W(1 + BI_W) = -\alpha_0^W\, dx \tag{28}$$

where $I_W \equiv I_W(x)$. Equation 28 can be integrated to yield (for the general case, B, C, D $\neq 0$):

$$\begin{aligned} \ln(I_W/I_W(0)] + [(C - 0) - D/B^2] \times \ln[1 + BI_W(0)] \\ + (D/B)[I_W - I_W(0)] = -\alpha_0^W x \end{aligned} \tag{29}$$

This transcendental equation can be solved graphically for a given incident write intensity $I_W(0)$, for each x.

Case a — Here we consider a molecular system in which the excited-state absorption in S and T manifolds can be neglected. However, we allow for population trapping in T_1; thus, we are actually examining propagation through a medium in which the ground-state population has been depleted (a typical saturable absorber behavior). In this case:

$$dI_W(1 + CI_W)/I_W = -\alpha_0^W\, dx \tag{30}$$

which is solved for propagating a length L in the medium to yield:

$$Y = \exp(-\alpha_0^W L)\exp[CI_W(0)(1 - Y)] \equiv G[Y,I_W(0)] \tag{31}$$

where

$$Y = I_W^{(L)}/I_W^{(0)} \tag{32}$$

Case b — Here we take into account excited-state absorption in addition to ground-state interaction. It is useful to rewrite Equation 29 in terms of G and Y:

$$Y = \exp(-\alpha_0^W L)\exp[(D/B)I_W(0)(1 - Y)] \times \{[1 + BI_W(0)]\}^{\delta} \equiv G(Y) \tag{33}$$

where δ is given by:

$$\delta = (D/B + B - C)/B \tag{34}$$

One can distinguish between two types of nonlinear absorption: (1) nonlinearly decreasing absorption when $\delta < 0$, mainly due to ground-state depletion; and (2) nonlinearly increasing absorption when $\delta > 0$, mainly due to large excited-state absorption coefficients. At very high intensity levels the increasing intensity-dependent absorption saturates.

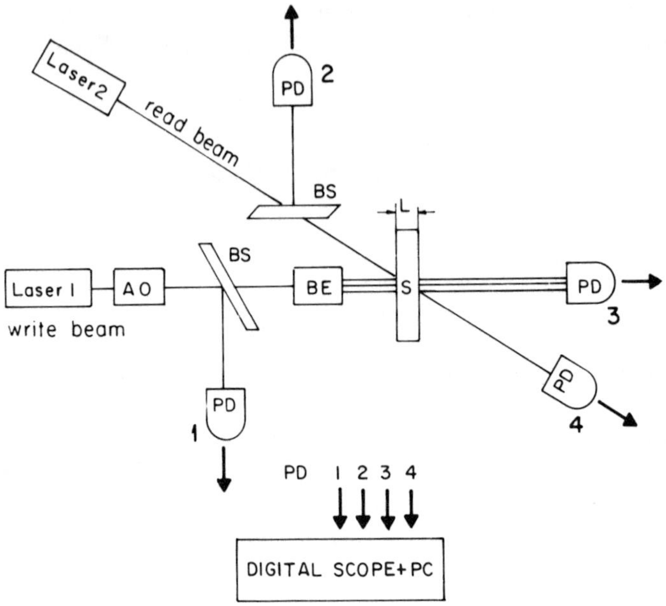

FIGURE 2. A typical setup for investigating write-read light beams interaction in a molecular SLM of thickness L: BS, beam splitter; BE, beam expander; AO, acoustooptic modulator; PD, photodiode; S, sample.

2. Propagation of the Read Beam

A molecular spatial light modulator can be created by interrogating the excited medium by a read beam. The geometric configuration of such a device is shown in Figure 2.

The SLM performance is determined by the propagation of the read light through the x stratified medium. The intensity of the read light $I_R(x)$ obeys Beer's law relation (for $I_R \ll I_W$):

$$\frac{dI_R(x)}{dx} = -\alpha_R(x)I_R(x) \tag{35}$$

where the absorption coefficient of the read light beam is given by the scalar product:

$$\alpha_R = \vec{N}[I_W(x)] \cdot \vec{\sigma}_R \tag{36}$$

and σ_R is the absorption cross-section vector at the reading light frequency:

$$\vec{\sigma}_R = [\sigma_{01}(\vec{R}),\sigma_{1n}(R),0,\sigma_{1n}(R),0] \tag{37}$$

Using the expression for $\vec{N}[I_W(x)]$,[10-12] we obtain:

$$\alpha_R = \alpha_0^R(1 + \beta I_w)/(1 + CI_W + DI_W^2) \tag{38}$$

where α_0^R and β are given in Table 2.

The solution of Equation 15 for a reading light beam that probes a length L of the medium is

$$I_R(L) = I_R(0)\exp\left(-\int_0^L \alpha_R \, dx\right) \tag{39}$$

for a read beam propagating in the direction of the writing beam and for incident read intensity $I_R(0)$, and:

$$I_R(L) = I_R(0)\exp\left(\int_L^0 \alpha_R \, dx\right) \tag{40}$$

for counterpropagating write and read beams. Equations 19 and 20 can be integrated numerically for a given Y(x), x, and $I_W(0)$. However, the solutions can be obtained explicitly from Equations 10 and 15, from which we can write:

$$\frac{dI_R}{I_R} = \frac{(\vec{N} \cdot \vec{\sigma}_R)}{(\vec{N} \cdot \vec{\sigma}^W)}\left(\frac{dI_W}{I_W}\right) = \left(\frac{\alpha_0^R}{\alpha_0^W}\right)\frac{(1 + BI_W)}{(1 + BI_W)}\left(\frac{dI_W}{I_W}\right) \tag{41}$$

Equation 41 can be integrated to yield the intensity transmission of the SLM for the read beam:

$$Z = I_R(L)/I_R(0) \tag{42}$$

as

$$Z = (Y\{[1 + BI_W(0)Y]/[1 + BI_W(0)]\}^{\gamma})^{\xi} \tag{43}$$

where the excess excited-state absorption for the read beam is given by:

$$\gamma = \beta/B - 1 \tag{44}$$

and ξ is the ground-state absorption cross-section ratio:

$$\xi \equiv \alpha_0^R/\alpha_W^0 = \sigma_{01}^s(R)/\sigma_{01}^s(W) \tag{45}$$

Equation 42 is the basis for the discussion of the molecular SLM performance. For counterpropagating write and read beams a similar expression is obtained, where the power ξ is replaced by $-\xi$.

3. Triplet SLMs

In this case information is written on the SLM by interaction with only the ground state, while the read beam interacts mainly with the excited states. In this case Z is obtained by integrating Equation 41 for B = 0 to yield:

$$Z = (Y)^{\xi}\exp[-\beta' I_W(0)(1 - Y)] \tag{46}$$

where β' is given in Table 2.

When the read beam interacts mainly with excited singlet states, much higher write intensities are needed than for read beam interactions with mainly triplet-state SLM.

The performance of an SLM is usually discussed by examining a few performance parameters.

a. Resolution

The spatial resolution in our case is solely determined by the external beam optics, since the microscopic interaction is with single molecules. We can thus estimate the resolution to be about 1 μm. Of course, this is true only for short-lived S_1 SLMs or for long-lived T_1 SLMs in rigid media such as polymer matrices, where diffusion during the SLM operation in negligible.

b. Contrast Ratio

The maximum contrast ratio can be defined as:

$$R = Z(I_W^{max})/Z(I_W^{min}) \tag{47}$$

For the SLM characterized by Equation 46, a normally open SLM, *R* is given by:

$$R = \exp[\sigma_{1n}^{T}(R)NL] \tag{48}$$

For the normally closed SLM *R* is given by:

$$R = \exp[\sigma_{01}^{s}(R)NL] \tag{49}$$

Thus, for both the normally closed and normally open SLMs, the contrast ratio depends exponentially on the length and on the absorption characteristics of the medium in either the ground or excited states. This feature is significant for the design of a molecular SLM.

c. Linearity and Sensitivity

The main drawback of a molecular SLM seems to be the lack of a wide linear range, i.e., a range for which Z varies linearly with I_W. Thus, we can only define an average sensitivity for the device:

$$S = \frac{\partial Z}{\partial I_W(O)} \tag{50}$$

d. Temporal Behavior

Our discussion pertains to steady-state conditions. For triplet-state SLM, it means a temporal response of the order of τ_{10}^T, a relatively slow SLM. It is, however, faster than a photochromic SLM.[7]

4. Transient Effects in Spatial Light Modulation

The treatment of SLM performance discussed in Section III.A was limited to steady-state behavior of the molecular SLM. For triplet SLMs it means working with either cw write and read light sources or with pulsed sources of duration $t \gg \tau_{10}^T$. There are situations where this is not the case, or that a specific application calls for an analysis of the transient behavior of the molecular SLM which will be observed for $t \ll \tau_{10}^T$.

Recently we have performed such an analysis,[13] treating the molecular absorbers of Figure 1a by assuming steady-state behavior for only levels S_n and T_n. The results for the molecular populations are

$$\vec{N} \equiv \vec{N}(I,x,t) = \frac{N'(S_0)}{K} \begin{pmatrix} K(1 + b_T E/b_s) \\ L_1 I(1 + b_T E/b_s) \\ L_1 L_2 I^2 (1 + b_T E/b_s), \\ L_1 k_{ISC} \tau_{10}^T I(1 - E) \\ L_1 L_3 I^2 \tau_{10}^T K_{ISC}(1 - E) \end{pmatrix} \tag{51}$$

where $N'(S_0)$ is the steady-state value of $N(S_0)$ given by Equation 20 and:

$$L_1 \equiv \sigma_{01}^S \tau_{10}^S \tag{52}$$

$$L_2 \equiv \sigma_{1n}^S \tau_{n1}^S \tag{53}$$

$$L_3 \equiv \sigma_{1n}^T \tau_{n1}^T \tag{54}$$

$$b_S \equiv K + L_1 I + L_1 L_2 I^2 \tag{55}$$

$$b_T \equiv L_1 k_{ISC}(1 + L_3 I)\tau_{10}^T I \tag{56}$$

$$E \equiv \exp[-(t/\tau_{10}^T)(1 + b/bs)] \tag{57}$$

The transient behavior for the transmission of the write beam for the molecular parameter characteristic of a triplet SLM is given by:

$$Y = \exp(-\alpha_0^W L)\exp\{M I_W(O)(1 - Y)[1 - \exp(-t/\tau_{10}^T)]\} \quad (58)$$

where

$$M = \sigma_{01}^S(W)\tau_{10}^S k_{ISC}\tau_{10}^T/K \quad (59)$$

and the corresponding transmission for the read beam is

$$Z = Y^{\xi}\exp\{-\beta'(1 - Y)I_W(O)[1 + \exp(-t/\tau_{10}^T)]\} \quad (60)$$

$$R(t) = \exp\{\sigma_{1n}^T(R)[1 + \exp - t/\tau_{10}^T]NL\} \quad (61)$$

which will reach its maximum value of:

$$R = \exp[2\sigma_{1n}^T(R)NL] \quad (62)$$

for $t \ll \tau_{10}^T$. Comparison with Equation 48 indicates that the transient contrast ratio can be as high as the square of its steady-state value. This behavior was verified in a recent experiment.[13] In addition, study of the temporal behavior of R (Equation 61) can be utilized to determine $\sigma_{1n}^T(W)$ values for a given molecular absorber.

5. Photochromic SLM

The above analysis pertains to dynamic photochromism based on triplet-triplet absorption. A related molecular system is that based on photochromic isomerization[14] described in Figure 1b. The simplification compared to the case of triplet SLM is in the neglect of S_n state and the replacement of T_1 and T_n by P_0 and P_1, the energy levels of the photoisomer. In addition τ_p, which is the equivalent of τ_{10}^T, is much longer than all other lifetimes or pulse duration. The kinetic analysis of the photochromic system level diagram (Figure 1b) yields the following results.[13]

The concentration of photoisomers:

$$N(P) = Nk\tau_p \frac{1 - \exp\{-t(1/\tau_p + kI_W\}}{1 + k\tau_p} \quad (63)$$

where

$$k = \sigma_{01}^S(W)\tau_{10}^S k_{ISO}/K' \quad (64)$$

and k_{ISO} is the rate constant for photoisomerization (Figure 1b) and $K' = 1 + \tau_{10}^S k_{ISO}$. For long-lived photoisomers we obtain:

$$N(P) = N(1 - \exp(-kI_w t)) \quad (65)$$

The nonlinear absorption coefficient of the photochromic system is obtained by calculating:

$$\alpha_W^P(I_W,t) = \vec{N}_p(I_W,t) \cdot \vec{\sigma}_p(W) \tag{66}$$

with the substitutions:

$$\vec{\sigma}_p(W) = (\sigma_{01}^S(W),0,0,\sigma_p(W),0) \tag{67}$$

and the population vector:

$$\vec{N}(I_W,t) = N\begin{pmatrix} \exp(-kI_Wt) \\ [KI_W\exp(-kI_Wt)]/k_{ISO} \\ 0 \\ 1 - \exp(I_Wkt) \\ \sigma_p(W)\tau_pI_W[1 - \exp(-kI_Wt)] \end{pmatrix} \tag{68}$$

to yield:

$$\alpha_W^P(t) = \alpha_0^P(W)\{[1 - \gamma_p^W]\exp(-kI_Wt) + \gamma_p^W\} \tag{69}$$

where

$$\gamma_p^W = \sigma_p(W)/\sigma_{01}^S(W) \tag{70}$$

Following the same procedure and approximation used in analyzing triplet SLM we find that for the photochromic system:

$$Y = \{\exp[-\alpha_0^P(W)L]\}\exp[kI_W(0)(1 - Y)t] \tag{71}$$

which for the steady-state operation reduces to:

$$Y = \exp[-\sigma_p(W)NL] \tag{72}$$

Similar to Equation 73 the read light absorption coefficient reads:

$$\alpha_R^P = \alpha_0^P(R)\{(1 - \gamma_p^R)\exp(-kI_Wt) + \gamma_p^R\} \tag{73}$$

where

$$\gamma_p^R = \sigma_p(R)/\sigma_{10}^S(R) \tag{74}$$

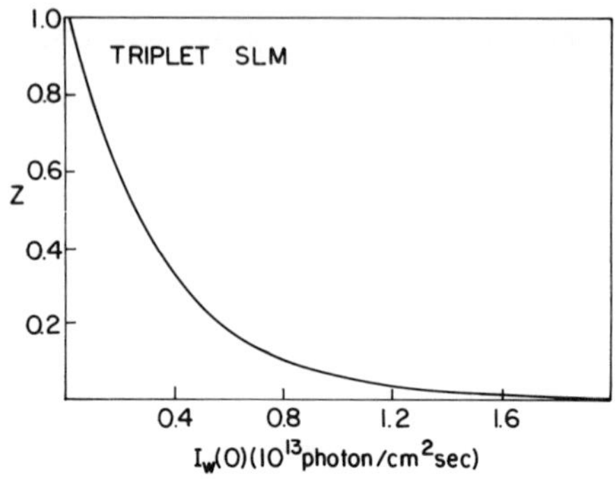

FIGURE 3. Characteristic performance intensity modulation curve of a triplet SLM with L = 10 μm, α_0^B = 60 cm^{-1}, α_0^W = 0.6 cm^{-1}, and τ_{10}^T = 100 μsec.

The short pulse excitation limit is

$$\alpha_R^P \rightarrow \alpha_0^P(R) \tag{75}$$

while the steady-state limit ($t \rightarrow \infty$) is

$$\alpha_R^P \rightarrow \sigma_p(R)N \tag{76}$$

The corresponding transmission characteristic of the photochromic SLM is

$$Z = Y^{\xi}\exp[\sigma_p(R)\tau_{10}^S k_{ISO}(1 - Y)I_W(0)t/K'] \tag{77}$$

which for the steady-state operation reduces to:

$$Z = \exp[-\sigma_p(R)NL] \tag{78}$$

From Equation 80 the steady-state contrast ratio is

$$R_p = \exp[\sigma_p(R)NL] \tag{79}$$

which is similar to that of a triplet SLM (Equation 48).

6. Numerical and Experimental Studies of Molecular SLMs

Numerical calculation of various types of molecular SLMs were performed. The best performance in terms of contrast ratio and other performance parameters was obtained for a triplet-state absorber SLM whose characteristic read light transmission curve is given by Equation 46. A typical curve calculated for a molecule with τ_{10}^T = 100 μsec, α_0^W = 60 cm^{-1} and α_0^R = 0.6 cm^{-1} for a path length of 10 μm is shown in Figure 3. The importance

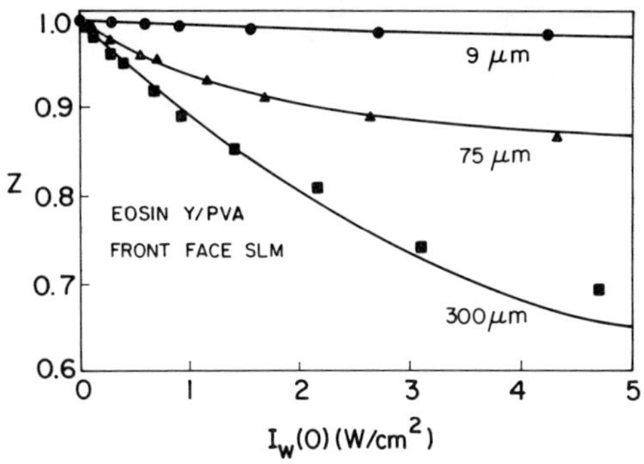

FIGURE 4. Normally open SLM experiments for a 1.1×10^{18} molecules per cubic centimeter eosin/PVA sample for three different sample thicknesses.

of this result is that it demonstrates the feasibility of creating a useful SLM provided high enough $\sigma_{1n}^{T}NL$ value can be realized.[12]

Attempts toward designing molecular SLMs were conducted. The choice was thin polymer eosin Y/doped films.[12] This particular substrate was used in related nonlinear optical molecular studies. Its molecular parameters promise high triplet-triplet absorption but not as those of phenosafranin.[15] The major drawback for an optimal SLM behavior for eosin Y is the degree of molecular aggregation observed for this molecule.[12] In previous studies of a related compound, disodiumfluorescein, aggregation[16] was shown to limit nonlinear optical response through triplet-triplet absorption while opening another, less efficient route based on dimer photodissociation.[17] While this latter route did not provide high enough optical nonlinearities for this molecule, it will be useful to design molecular dimers which can be photodissociated in a reversible fashion to a highly absorbing monomer, thus creating a new type of nonlinear photochromic material.[17]

Aggregation of eosin Y puts a very low limit on the molecular concentration N that can be used thus reducing $\sigma_{1n}^{T}NL$. High contrast ratios were obtained only by increasing L. This is demonstrated in Figures 4 and 5 where the performance of the eosin Y SLM is shown. For a front face SLM configuration (Figure 2) increasing L provides a poor solution for achieving high contrast ratios. However, the alternative of transient waveguide configuration (Figure 6), in which a transient waveguide is created by a writing light that is then probed by a read beam traveling along the waveguide, provides a novel type of molecular nonlinear switch where high contrast ratio is achieved at low concentration and low write beam intensities. Its performance is shown in Figure 7; however, the possibility of using this SLM for optical information processing, using a large area of pumped substrate, is lost.

The alternative for creating large R values is by designing novel polymer materials in which N can be increased without giving rise to destructive

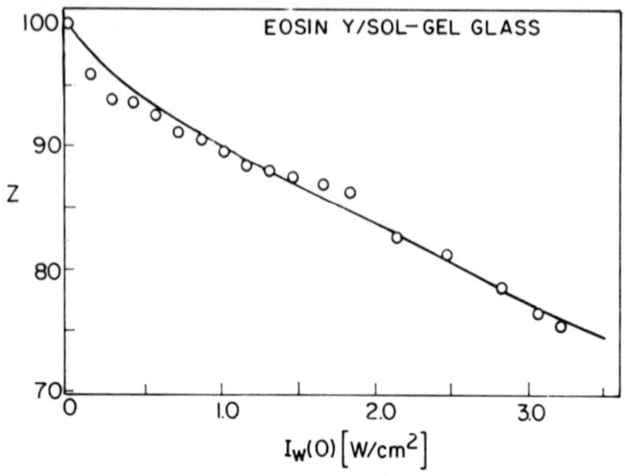

FIGURE 5. Normally open SLM performance for eosin Y/sol-gel glass sample of 1-μm thickness.

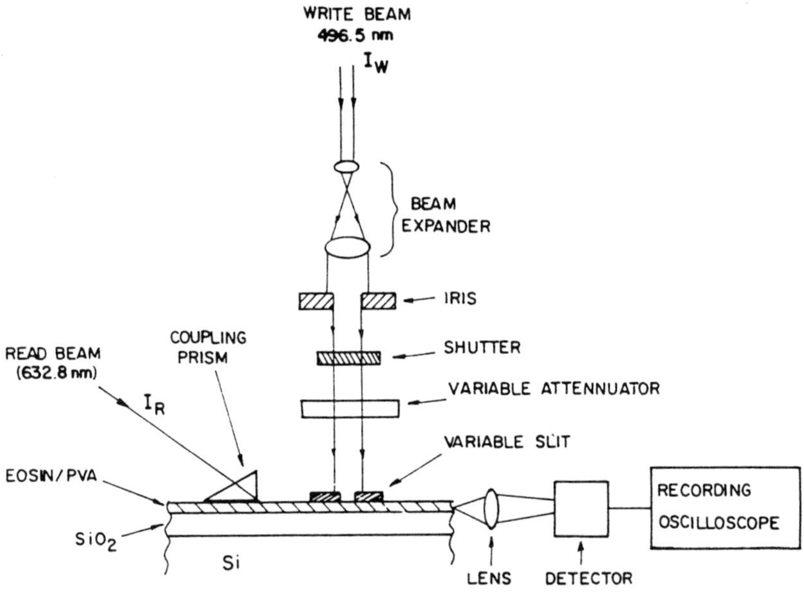

FIGURE 6. Typical experimental setup for transient waveguide switching experiments.

aggregation effects. Such polymer schemes were demonstrated recently by ion exchange techniques utilizing the ionic nature of most dyes.[18]

Another approach is the use of sol-gel glasses as an alternative host of doped dyes in which aggregation is limited by the sol-gel matrix structure.[19,20]

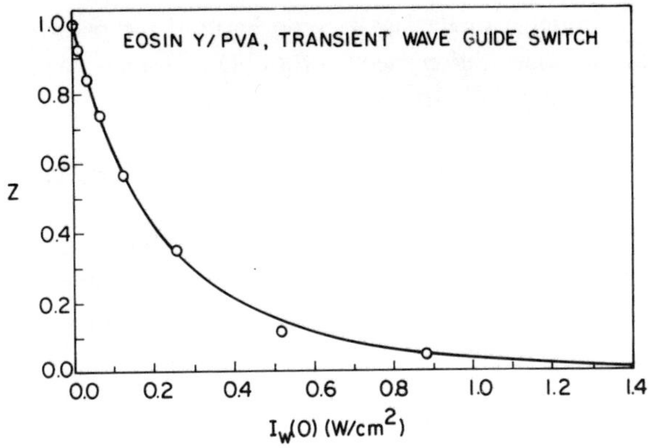

FIGURE 7. Transient waveguide all optical switch performance for 1.1×10^{18} molecules per cubic centimeter eosin/PVA sample of 1-cm path length.

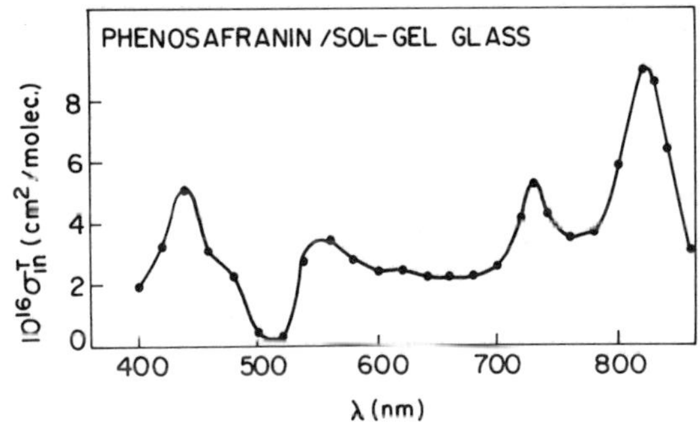

FIGURE 8. Triplet-triplet absorption spectrum of phenosafranin embedded in sol-gel glass matrix.

A useful combination of phenosafranin embedded in sol-gel matrix was recently shown to have promising properties. In particular, the large triplet-triplet absorption of phenosafranin[15] at the 600 to 850 nm region is a very attractive feature (Figure 8).[21]

A related SLM is that based on a photochromic system (Equation 79). Kirkby et al.[7] examined a photochromic molecule in an SLM configuration. In this particular system they obtained a contrast ratio of R = 12 which is comparable to that obtained for eosin Y. A much better performance was demonstrated recently in the bacteriorhodopsin cycle-based system. The suitability of this SLM for optical data processing was demonstrated.[22]

Further search for molecular systems suitable for spatial light modulation is needed. The ideal system should be based either on high triplet-triplet absorption[23] or a photochromic isomer absorbing at the 700- to 900-nm spectral region so that suitable diode lasers could be used for read/write functions. High molecular concentration, needed for obtaining high contrast ratios, can be achieved by using polymer ion exchange schemes.[18]

B. SYSTEMS WITH OPTICAL FEEDBACK: OPTICAL BISTABILITY

Optical bistability is characterized by two different light transmission states of an optical system for a given input light intensity.[24-26] A hysteresis loop of output (I_{out}) vs. input (I_{in}) light intensity, indicating a kind of memory to the system, is a manifestation of bistable response. In order to observe optical bistability, a nonlinear optical medium and an optical feedback are required. The optical nonlinearity is achieved by employing suitable nonlinear optical materials either dispersive nonlinearity, such as in materials which have large Kerr constants, or absorptive nonlinearity. The optical feedback is provided by the choice of a suitable optical resonator, such as a Fabry-Perot or ring resonator. It is the unique combination of a nonlinear optical material inside an optical resonator which determines the details of the bistable response. Optical bistability is significant in optical computing and in optical data processing. Optical bistable response means that one has a kind of optical switch which also can be used as a logic element. The dynamics of this switch depend to a large extent on the optical properties of the intracavity material. A nonlinear absorbing medium in an optical resonator was the first configuration for which the existence of optical bistability was theoretically predicted by Szoke et al.,[26] Seidel,[27] and McCall.[28]

Although difficult to implement experimentally, absorptive bistability has been a major subject of theoretical research involving elucidation of the mechanisms and characterization of the relevant microscopic phenomena. The main part of this theoretical work, initiated by Bonifacio and Lugiato,[29] was conducted for a nonlinear medium consisting of an ensemble of two-level systems, thus exhibiting saturation of resonant absorption. In this context, analysis of the steady-state and temporal behavior of the stability conditions were carried out,[30-32] for the most part using the mean-field approximation, which ignores propagation effects. Recent efforts concentrated on media exhibiting dynamically increasing absorption.[33]

The eikonal approximation was introduced in order to provide a rather simple analytical tool which has a predictive power as to the best choice of material-resonator combination.[34,35] In addition, we have analyzed the steady-state characteristics of absorptive bistability assuming particular microscopic models for the optical nonlinearity.[36] These studies established the relations between molecular parameters and the conditions for observing optical bistability. Since optical bistability in an organic dye has been observed,[37,38] we have also considered nonlinear molecular absorbers such as organic laser dyes.[39] The nonlinearity originates from unique intensity dependence of the complex index of refraction, manifested also in the photoquenching of molecular fluorescence.[40] For molecular materials, such as dyes, it connects the photophysical properties with the bistable response for a given cavity.[36]

The nonlinear complex eikonal approximation is a procedure for the treatment of light propagation in media characterized by a complex nonlinear refractive index, $\eta(I) = \eta^D(I) + i\eta^A(I)$. As indicated, in general, both the dispersive and the absorptive parts of the refractive index depend on the local intensity I(x). The eikonal approximation is expressed in terms of the phase accumulated upon propagation through a distance x in the medium:[35]

$$\phi(x) = (2\pi/\lambda_0) \int_0^x \eta[I(x')]dx \tag{80}$$

The local intensity depends on the local accumulated phase; it also depends, as a consequence of the boundary conditions, on the phase $\phi(L)$ accumulated along the medium length L. This procedure results in the integral equation:[34,35]

$$\phi(x) = (2\pi/\lambda_0) \int_0^x \eta\{I[x',\phi(x'),\phi]\}dx' \tag{81}$$

where

$$\begin{aligned} \phi(x) &= \phi_D(x) + (i/2)\phi_A(x) \\ \phi &= \phi_D + (i/2)\phi_A \equiv \phi(L) \end{aligned} \tag{82}$$

and λ_0 is the vacuum wavelength. $\phi_D(x)$ and $\phi_A(x)$ denote the dispersive and absorptive contributions to $\phi(x)$, respectively. The factor of $^1/_2$ is introduced to follow common conventions.

It will be convenient to express the absorptive part of the refractive index in terms of the absorption coefficient $\alpha(I)$:

$$\eta^A(I) = \lambda_0\alpha(I)/4\pi \tag{83}$$

The procedure for obtaining the transmission characteristics I_{out} vs. I_{in} of a particular optical configuration consists of the following steps:[34-36]

1. The appropriate form for the nonlinear index of refraction $\eta(I)$ and the boundary conditions are specified. The latter, in turn, determine the form of $I[x, \phi(x), \phi(L)]$.
2. The integral Equation 81 for $\phi(x)$ is solved or alternatively its differential form is used:

$$d\phi(x)/\eta\{I[x,\phi(x),\phi]\} = (2\pi/\lambda_0)dx \tag{84}$$

which is solved by separation of variables and integration.

3. By imposing self-consistency $\phi(L)$ is obtained.
4. I_{out} vs. I_{in} is calculated.

In many situations either the absorptive or the dispersive nonlinearity can be ignored. For a purely dispersive nonlinearity,[34-36] the following equation is obtained:

$$d\phi_D(x)/\eta^D\{I[x,\phi_D(x),\phi_D]\} = (2\pi/\lambda_0)dx \tag{85}$$

whereas in the purely absorptive case:[36]

$$d\phi_A/\alpha\{I[x,\phi_A(x),\phi_A]\} = dx \tag{86}$$

We have demonstrated the applicability of molecular systems for obtaining optical bistability. Although our discussion was limited to ring resonators, the main features are maintained in other configurations such as the Fabry-Perot resonator.[34,35]

The main advantages in utilizing molecular systems as media for bistable optical devices lie in the variety of molecular nonlinear parameters that are involved in inducing optical bistability. Different switching patterns can be obtained, depending on the particular switching parameter, i.e., switching by intensity variation, absorption coefficient scan (spectral scan), medium length, etc.

Molecular systems provide the first nonlinear absorptive practical media for observing optical bistability by incoherent light excitation.[39] Since the optical bistability characteristics depend strongly on the particular molecular parameters, the observed phenomena can be utilized as a novel means for photophysical studies. Numerical analysis shows that low excited-state absorption coefficients can be measured by following the variation in the bistable loop size (the difference between up and down switching thresholds intensities). Frequency scan of a molecular bistable device may be utilized as a sharp-edge interference filter of very low transmittance (Figure 9). In addition, information on index of refraction associated with a particular molecular level can be obtained from optical bistability studies.[39]

Experimental tests for these numerical predictions were performed on disodium fluorescein doped PMMA samples.[10,41] Depending on the pumping wavelength, optical bistability changed from the normal clockwise loop characteristic of saturable absorber behavior (Figure 10) to anticlockwise loop characteristic of increasing absorption bistability.[33] This change in behavior is a manifestation of the molecular nonlinearity which results in the optical bistability spectrum shown in Figure 11. More recent results[41] show the same effect in eosin Y doped sol-gel glass samples (Figures 12 and 13). Similar results were recently observed by Sojka and Papanek in aqueous solutions of toluidine blue,[42] where dimer-monomer equilibrium is the source for optical

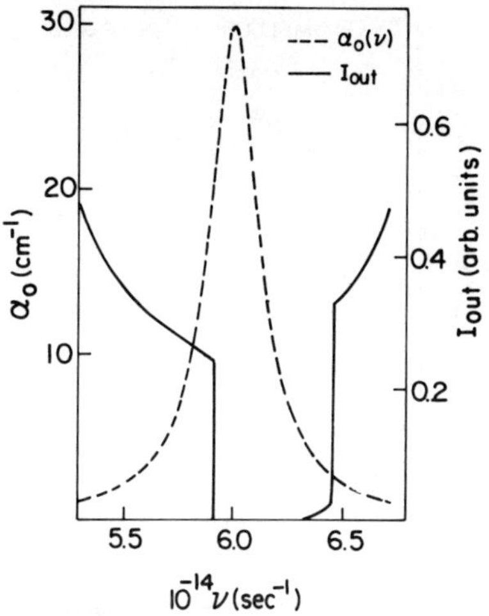

FIGURE 9. Frequency scan of the output intensity of a bistable optical element based on nonlinear optical absorption together with the absorption spectrum $\alpha_0(\nu)$.

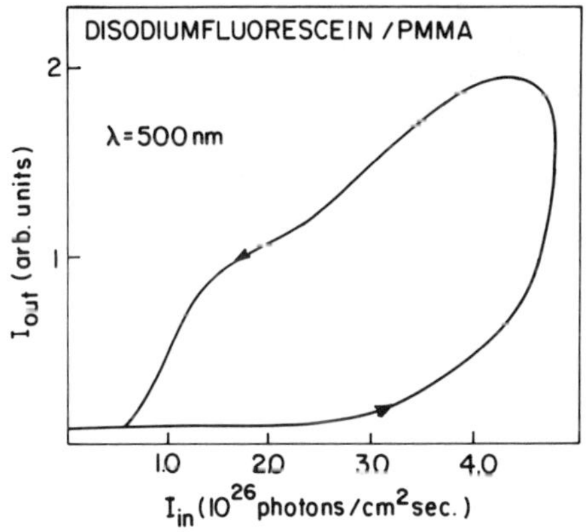

FIGURE 10. Optical bistability exhibited for disodium fluorescein/PMMA thin film (1 μm) sample pumped at 500 nm; a typical saturable absorber performance.

nonlinearity in this system.[17] All of these results indicate that an optical bistable element based on nonlinear molecular absorption is feasible.

IV. RELATED NONLINEAR OPTICAL PHENOMENA

The $\chi^{(3)}$-type optical nonlinearities which give rise to the observed phenomena of spatial light modulation and optical bistability can be also manifested in other forms.

Of particular interest are four-wave mixing (FWM).[43] In this case coupling between material excitations and the light waves are utilized. The material

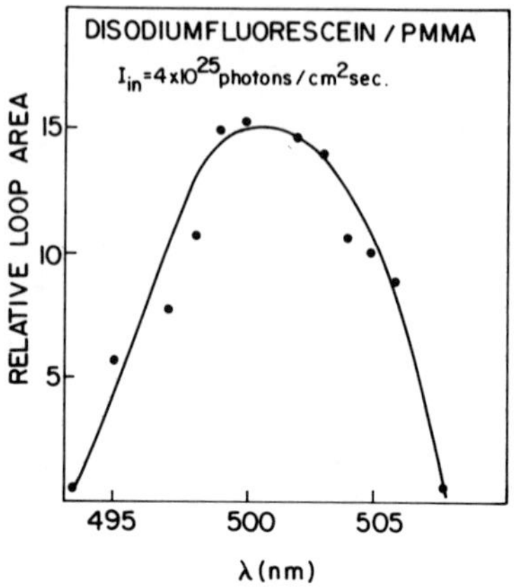

FIGURE 11. Optical bistability spectrum for disodium fluorescein/PMMA thin film (1 μm) sample indicating maximum efficiency at the saturable absorber wavelength 500 nm with diminishing efficiency at other wavelengths where increasing absorption is the dominant nonlinear optical response.

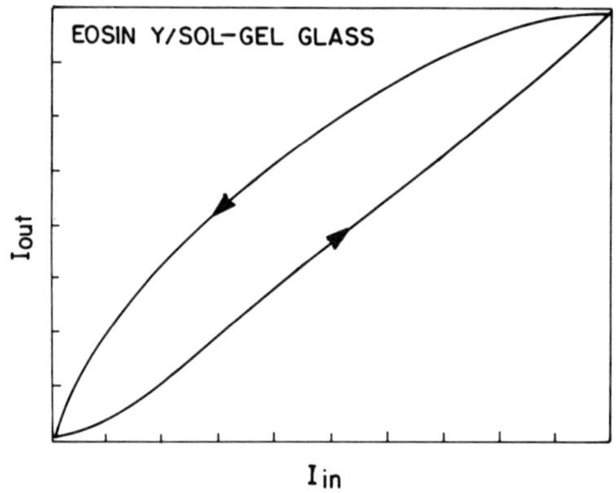

FIGURE 12. Optical bistability exhibited for eosin Y/sol-gel glass thin film.

excitations appear in the form of dynamic grading[44] that can coherently scatter incoming radiation. The material excitation can then be followed by probing the dynamic gratings. When all light waves are of the same frequency, conditions are set for degenerate four-wave mixing (DFWM)[43,45,46] that can be used for amplified reflection and phase conjugation.[47] Propagation effects, which are especially important in the case of saturable absorbers, were discussed by Boyd and co-workers.[48,49] Transient effects[50] and polarization[51] effects of DFWM in dye media were discussed and demonstrated experimentally.

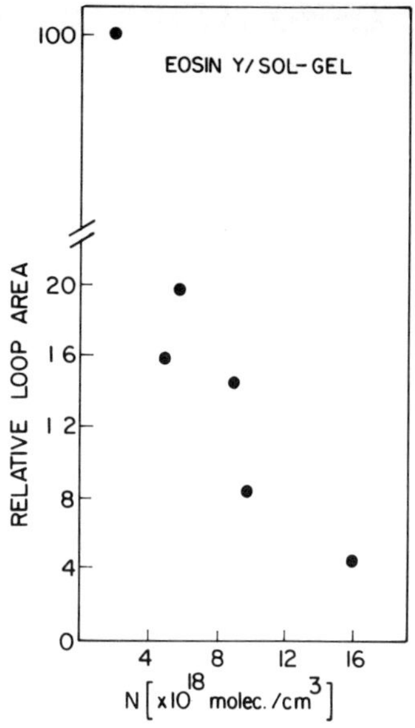

FIGURE 13. Efficiency of the optical bistable response of eosin Y/sol-gel glass sample as function of the sample concentration.

This type of third order mixing technique can be used to generate phase-conjugated waves which are of great interest for adaptive optics and real time holography applications. Both absorptive and dispersive contribution of $\chi^{(3)}$ can be utilized in these cases.[46,52]

Low power phase conjugation was demonstrated in saturable absorbers, such as eosin Y, embedded in gelatin films[53,54] and other dyes.[55] Similar behavior was observed in PMMA doped samples.[56] The same and similar nonlinear optical materials were also utilized in real time holography studies and other optical data processing procedures.[57-59]

V. CONCLUDING REMARKS

In this review chapter we have discussed the general requirements needed to realize nonlinear optical switching by molecular media.[60] Absorptive, in addition to dispersive optical nonlinearity,[61] based on dynamic population changes in organic dyes associated with triplet-triplet absorption can be utilized to construct useful optical switches or optical bistable elements. Alternatively photochromic systems can be used in a similar fashion.

Photoaddressed spatial light modulators can be constructed by using suitable nonlinear optical polymers,[62] or by utilizing induced birefringence in organic dyes.[63] The importance of obtaining highly absorbing materials either by novel synthetic methods[64] or by incorporating dye chromophores in a polymer[18] was emphasized.

ACKNOWLEDGMENT

This research was supported by the Fund for Promotion of Research at the Technion.

REFERENCES

1. **Shen, Y. R.,** *The Principles of Nonlinear Optics,* John Wiley & Sons, New York, 1984.
2. **Yu, F. T. S.,** *Optical Information Processing,* John Wiley & Sons, New York, 1983.
3. **Casasent, D.,** Spatial light modulators, *Appl. Opt.,* 14, 2445, 1979; Coherent light values, *Appl. Opt. Opt. Eng.,* 5, 143, 1980.
4. **Tanguay, A. R., Jr.,** Materials requirement for optical processing and computing devices, *Opt. Eng.,* 24, 2, 1985.
5. **Chemla, D. S. and Zyss, J.,** *Nonlinear Optical Properties of Organic Molecules and Crystals,* Academic Press, New York, 1987.
6. **Williams, D. J., Ed.,** *Nonlinear Optical Properties of Organic and Polymeric Materials,* ACS Symposium Series 233, ASC, Washington, D.C., 1983.
7. **Kirkby, C. J. G., Cush, R., and Bennion, I.,** Optical nonlinearity and bistability in organic photochromic thin films, *Opt. Commun.,* 56, 288, 1985; **Kirkby, C. J. G. and Bennion, I.,** Organic photochromics for spatial light light modulation, *IEEE Proc.,* 133, 98, 1986.
8. **Kramer, M. A. and Tompkin, W. R.,** Nonlinear-optical interactions in fluorescein doped boric acid glass, *Phys. Rev. A,* 34, 2026, 1986.
9. **Speiser, S. and Chisena, F. L.,** Optical bistability in fluorescein dyes, *Appl. Phys. B,* 45, 137, 1988; Optical bistability in dye molecules, *J. Chem. Phys.,* 89, 7259, 1988; Optical bistability in dimer-monomer dye system, *SPIE,* 824, 144, 1988.
10. **Speiser, S. and Orenstein, M.,** Spatial light modulation via optically induced absorption changes in molecules, *Appl. Opt.,* 27, 2944, 1988.
11. **Speiser, S., Dansker, D., and Orenstein, M.,** Spatial light modulation by nonlinear absorbers, *J. Appl. Phys.,* 66, 61, 1989.
12. **Beeson, K. W., Yardley, J. T., and Speiser, S.,** Utilization of nonlinear optical absorption in eosin Y for all optical switching, *Mol. Eng.,* 1, 1, 1991.
13. **Dantsker, D. and Speiser, S.,** Time dependent spatial light modulation by nonlinear molecular absorbers, *Nonl. Opt.,* in press.
14. **Dürr, H. and Bonas-Laurent, H.,** *Photochromism, Molecules and Systems,* Elsevier, Amsterdam, 1990.
15. **Gopidas, K. and Kamat, P. V.,** Photophysics and photochemistry of phenosafranin dye in aqueous and acetonitrile solutions, *J. Photochem. Photobiol. A. Chem.,* 48, 291, 1989.
16. **Lopez-Arbeloa, I.,** Aggregation of halofluorescein dyes, *Dyes Pigments,* 4, 213, 1983.
17. **Speiser, S., Houlding, V. H., and Yardley, J. T.,** Nonlinear optical properties of organic dye dimermonomer systems, *Appl. Phys. B,* 45, 237, 1988.
18. **Tomiyama, H., Okada, S., Matsuda, H., and Nakanishi, H.,** Novel third order nonlinear optical materials composed of ionic polymers and chromophores, *SPIE,* 1337, 170, 1990.
19. **Ulrich, D. R.,** Prospects of sol-gel processes, *J. Non Cryst. Solids,* 100, 174, 1988.
20. **Avnir, D., Kaufman, V. R., and Reisfeld, R.,** Organic fluorescent dyes trapped in silica and silicatitania thin films by the sol-gel method. Photophysical, film and cage properties, *J. Non Cryst. Solids,* 74, 395, 1985.

21. **Speiser, S.,** Nonlinear optical properties of phenosafranin doped substrates, *SPIE,* 1559, 238, 1990.
22. **Thoma, R., Hampp, N., Bruchle, C., and Oesterhet, D.,** Bacteriorhodopsin films as SLMs for nonlinear optical filtering, *Opt. Lett.,* 16, 651, 1991.
23. **Carmichael, I. and Hug, G. L.,** Triplet-triplet absorption spectra of organic molecules in condensed phases, *J. Phys. Chem. Ref. Data,* 15, 1, 1986.
24. **Goldstone, J. A.,** Optical bistability, in *Laser Handbook,* Vol. 4, Stitch, M. L. and Bass, M., Eds., North-Holland, Amsterdam, 1985.
25. **Gibbs, H. M.,** *Optical Bistability: Controlling Light with Light,* Academic Press, New York, 1985.
26. **Szoke, A., Dannell, V., Goldhar, J., and Kurnit, N. A.,** Bistable optical element and its applications, *Appl. Phys. Lett.,* 15, 376, 1969.
27. **Seidel, H.,** U.S. Patent 3,610,731, 1969.
28. **McCall, S. L.,** Instabilities in continuous-wave light propagation in absorbing media, *Phys. Rev. A,* 9, 1515, 1974.
29. **Bonifacio, R. and Lugiato, L. A.,** Optical bistability and cooperative effects in resonance fluorescence, *Phys. Rev. A,* 18, 1129, 1978.
30. **Bischofberger, T. and Shen, Y. R.,** Theoretical and experimental study of the dynamic behavior of a nonlinear Fabry-Perot interferometer, *Phys. Rev. A,* 19, 1169, 1979.
31. **Smith, S. D.,** An introduction to optically bistable devices and photonic logic, *Philos. Trans. R. Soc. London, Ser. A,* 313, 195, 1984.
32. **Goldstone, J. A. and Garmire, E. M.,** On the dynamic response of nonlinear Fabry-Perot interferometers, *IEEE J. Quantum Electron.,* QE-17, 366, 1981.
33. **Miller, D. A.,** Optical bistability and differential gain resulting from absorption increasing with excitation, *J. Opt. Soc. Am. B,* 1, 857, 1984.
34. **Orenstein, M., Speiser, S., and Katriel, J.,** An eikonal approximation for nonlinear resonators exhibiting bistability, *Opt. Commun.,* 48, 367, 1984.
35. **Orenstein, M., Speiser, S., and Katriel, J.,** A general eikonal treatment of coupled dispersively nonlinear resonators exhibiting optical multistability, *IEEE J. Quantum Electron.,* QE-21, 1513, 1985.
36. **Orenstein, M., Katriel, J., and Speiser, S.,** Nonlinear complex eikonal approximation: optical bistability in absorbing media, *Phys. Rev. A,* 35, 1192, 1987.
37. **Zhu, Z. F. and Garmire, E. M.,** Optical bistability in BDN dye, *IEEE J. Quantum Electron.,* 19, 1495, 1983.
38. **Rushford, M. C., Gibbs, H. M., Jewell, J. L., Peyghambarian, N., Weinberger, D. A., and Li, C. F.,** Observation of thermal optical bistability, crosstalk, regenerative pulsations, and external switch-off in a simple dye-filled etalon, *Optical Bistability 2,* Bowden, C. M. and McCall, S. L., Eds., Plenum Press, New York, 1983, 345.
39. **Orenstein, M., Katriel, J., and Speiser, S.,** Optical bistability in molecular systems exhibiting nonlinear absorption, *Phys. Rev. A,* 35, 2175, 1987.
40. **Speiser, S. and Shakkour, N.,** Photoquenching parameters for commonly used laser dyes, *Appl. Phys. B,* 38, 191, 1985 (and references therein).
41. **Speiser, S.,** Model systems for optoelectronic devices based on nonlinear molecular absorption, *SPIE,* 1560, 434, 1991.
42. **Sojka, J. and Papanek, J.,** Increasing absorption optical bistability in highly concentrated dye solutions, *Opt. Commun.,* 85, 241, 1991.
43. **Shen, Y. R.,** Basic consideration of four wave-mixing and dynamic gratings, *IEEE J. Quantum Electron.,* QE-22, 1196, 1986.
44. **Eichler, H. J., Günter, P., and Pohl, D. W.,** *Laser Induced Dynamics Gratings,* Springer-Verlag, Berlin, 1986.
45. **Fayer, M. D.,** Dynamic of molecules in condensed phases: picosecond holographic grating experiments, *Annu. Rev. Phys. Chem.,* 33, 63, 1982.

46. **Abrams, R. L. and Lind, R. C.,** Degenerate four-wave mixing in absorbing media, *Opt. Lett.,* 2, 94, 1978.
47. **Yariv, A. and Pepper, D. M.,** Amplified reflection, phase conjugation, and oscillation in degenerate four-wave mixing, *Opt. Lett.,* 1, 16, 1977.
48. **Gruneisen, M. T., Gaeta, A. L., and Boyd, R. W.,** Exact theory of pump-wave propagation and its effect on degenerate four-wave mixing in saturable-absorbing media, *J. Opt. Soc. Am. B,* 2, 1117, 1985.
49. **Gaeta, A. L., Gruneisen, M. T., and Boyd, R. W.,** Theory of degenerate four-wave mixing in saturable absorbing media with the inclusion of pump propagation effects, *IEEE J. Quantum Electron.,* QE-22, 1095, 1986.
50. **Fujiwara, H. and Nakagawa, K.,** Transient phase conjugation by degenerate four-wave mixing in saturable dyes, *J. Opt. Soc. Am. B,* 4, 121, 1987.
51. **Tompkin, W. R., Malcurt, M. S., Boyd, R. W., and Sipe, J. E.,** Polarization of phase conjugation by degenerate four-wave mixing in a medium or rigidly held dye molecules, *J. Opt. Soc. Am. B,* 4, 757, 1989.
52. **Curo, R. G. and Gower, M. C.,** Phase conjugation by degenerate four-wave mixing in absorbing media, *IEEE J. Quantum Electron.,* QE-18, 1376, 1982.
53. **Silberberg, Y. and Bar-Joseph, I.,** Low power phase conjugation in thin films of saturable absorbers, *Opt. Commun.,* 39, 265, 1981.
54. **Nakaguwa, K. and Fujiwara, H.,** Real-time and double-exposure phase-conjugate interferometers using eosin-doped gelatin film, *Opt. Commun.,* 70, 73, 1989.
55. **Moses, E. I. and Wu, F. Y.,** Amplification and phase conjugation by degenerate four-wave mixing in a saturable absorber, *Opt. Lett.,* 5, 64, 1980.
56. **Todorov, T., Nikolova, L., Tomova, N., and Dragostinova, A.,** Photochromism and dynamic holographic recording in a rigid solution of fluorescein, *Opt. Quantum Electron.,* 13, 209, 1981.
57. **Bar-Joseph, I. and Silbergerg, Y.,** Real time holography through triplet-state absorption in organic dyes, *Opt. Commun.,* 41, 455, 1982.
58. **Fujiwara, H. and Nakagawa, K.,** Phase conjugation in fluorescein film by degenerate four-wave mixing and holographic process, *Opt. Commun.,* 55, 386, 1985.
59. **Moosad, K. P. B., Abdul Rasheed, T. M., and Nampoori, V. P. N.,** Optical phase conjugation in dyes embedded in polymer films, *Opt. Eng.,* 29, 47, 1990.
60. **Boyd, G. T.,** Application requirements for nonlinear-optical devices and the status of organic materials, *J. Opt. Soc. Am. B,* 4, 685, 1989.
61. **Chang, T. Y.,** Fast self-induced refractive index changes in optical media: a survey, *Opt. Eng.,* 20, 220, 1981.
62. **Stamatoff, J., Buckley, A., Swanson, G., Gacusan, L., Lytel, R., and Lipscomb, G. F.,** Nonlinear optical polymers for photo-addressed spatial light modulators, *SPIE,* 1150, 2, 1990.
63. **Montecchi, M., Settembre, M., and Romangoli, M.,** Optically induced birefringence and dichroism in rigidly held dye molecules, *J. Opt. Soc. Am. B,* 5, 2357, 1988.
64. **Fabian, J. and Zahradrik, R.,** The search for highly colored organic compounds, *Angew. Chem. Int. Ed. Engl.,* 28, 677, 1989.

Chapter 5

CLASSICAL ASPECTS OF QUANTUM TUNNELING

J. Gonzalo Muga, Rafael Sala, and Héctor Cruz

TABLE OF CONTENTS

0-8493-8061-8/93/$0.00 + $.50

I. INTRODUCTION

Tunneling is frequently considered one of the quintessential quantum phenomena. However, classical mechanics is useful for describing the tunnel effect, not only by means of approximate semiclassical methods, but also as a source of formal structures that provide exact quantum results, and help to visualize the dynamics. This chapter reviews recent work, mainly by one of the authors and collaborators, where tunneling and in general the quantum passage through a potential barrier are examined with the aid of classical mechanics.

Time-dependent tunneling under some conditions is amenable to *purely classical* explanation.[1] More generally, classical concepts like the trajectory serve after proper modification as a formal framework to reproduce quantum transmittance and wave function exactly.[2,3] This study is not only of conceptual interest: classical mechanics is usually easier to deal with than quantum mechanics from the computational point of view; thus it is useful to know when and how classical methods can provide a reliable answer. The language of classical statistical mechanics (phase space) is very useful for a variety of quantum mechanical applications. In particular, it is a natural framework in which to examine the asymptotic behavior of a quantum time-dependent or stationary collision,[4] and gives a clue for solving the ''tunneling time problem''. In any case, the structure of classical mechanics is rich enough to inspire new quantum methods, or semiclassical approximations, and definitely helps to achieve a better understanding of quantum physics.

This is *not* a review on semiclassical methods, or at least not on the usual meaning of the word ''semiclassical''. Specifically, the WKB method, the semiclassical S-matrix, and its variants are not examined here. The reader may find excellent and comprehensive discussions on these issues elsewhere. One more warning: for many applications, in particular for semiconductor devices, one dimensional models are appropriate and the discussion is limited to this simple case.

The text is organized as follows: in Section II the phase space formulation of scattering is used for obtaining formal transmittance expressions. It is shown how the transmittance is not, in general, the (properly weighted) average of the stationary transmittances as it is sometimes assumed. This average is just an approximation of the exact result. The approximate expression is, however, formally valid in classical mechanics, and its limits of validity are actually rather broad. Section III is devoted to the comparison between time-dependent quantum and classical statistical behavior. In Section IV the trajectory concept is extended to the quantum case for obtaining exact quantum wave functions with classical-like methods for stationary scattering. Section V deals with a phase space tunneling time theory inspired in classical reasoning. The chapter ends with a discussion and two Appendices on technical questions.

II. WAVE PACKET TRANSMITTANCE

Classical quantum comparisons are, in general, better implemented at the level of phase space representatives of the quantum state and the corresponding classical statistical ensembles. In this section, the transmittance is obtained with the phase space formulation of scattering,[5–7] in particular by means of a formalism introduced by Snider[6] and later developed by one of the authors and Snider.[4] Following Reference 4 we shall first discuss the density operator and the Wigner equivalent description of the collision. In order to find the transmittances for wave packets it is useful to obtain the asymptotic flux and transmission coefficients in the stationary case. These coefficients are, of course, known, but the mehtod of derivation is novel and serves to introduce the phase space technique.

In a time-dependent collision the density operator $\rho(t)$ is related to the incoming ''free'' density operator $\wp^{in}(t)$ through the Möller superoperator:

$$\rho(t) = \Omega_L \wp^{in}(t) \tag{1}$$

The definition of various scattering operators and superoperators is given in Appendix A. By using momentum resolutions of the identity, the incoming density operator is decomposed into its frequency components, and the time-dependent interacting density operator can then be given by:

$$\rho(t) = \lim_{\eta \to 0} \iint dp' dp'' e^{-i\omega_{p'p''}t} \Omega_L (\omega_{p'p''} + i\eta) |p'\rangle \wp^{in}_{p'p''}(0) \langle p''| \tag{2}$$

Here $\omega_{p'p''} = (p'^2 - p''^2)/(2m\hbar)$, and the assumed normalization is $\langle p''|p'\rangle = \delta(p'' - p')$. The correspondiong Wigner function is

$$W(x,p,t) \equiv \frac{1}{2\pi} \int e^{isx} \langle p + s\hbar/2 | \rho(t) | p - s\hbar/2 \rangle \, ds \tag{3}$$

In terms of this function, the flux at time t and position x is given, as in classical statistical mechanics, by the integral:

$$\begin{aligned} J(x,t) &= \int \frac{p}{m} W(x,p,t) dp \\ &= \int \frac{p}{m} dp \iint dp' dp'' W_{p'p''}(x,p) \wp^{in}_{p'p''}(0) e^{-i\omega_{p'p''}t} \end{aligned} \tag{4}$$

where $W_{p'p''}(x,p)$ is the phase space representation of the frequency component:

$$\lim_{\eta \to 0} \rho_{\omega p'p'',\eta} \equiv \lim_{\eta \to 0} \Omega_L(\omega_{p'p''} + i\eta) |p'\rangle\langle p''| \tag{5}$$

These frequency components of the density operator obey a Lippmann-Schwinger type of equation (used here is a shorthand notation $\wp^{\text{in}}_{\omega p'p''} \equiv |p'\rangle\langle p''|$ for a free motion frequency component. $\wp^{\text{in}}_{\omega p'p''}$ is an "operator", and should not be confused with the scalar amplitude $\wp^{\text{in}}_{p'p''}(0)$ of the initial state, see Equation 2):

$$\rho_{\omega,\eta} = \wp^{\text{in}}_{\omega} + (\omega + \mathcal{L}_0 + i\eta)^{-1}\,\mathcal{T}(\omega + i\eta)\,\wp^{\text{in}}_{\omega} \tag{6}$$

where $\mathcal{T}$ is the transition superoperator (see Appendix A). The Wigner function of the second term, W^{sc}_{ω}, (the scattered part) is given by:

$$W^{sc}_{\omega}(x,p) = (2\pi)^{-1}\int ds e^{isx}\,\langle p + s\hbar/2|(\omega - \mathcal{L}_0 + i\eta)^{-1}\,\mathcal{T}(\omega + i\eta)\,\wp^{\text{in}}_{\omega}|p - s\hbar/2\rangle$$

$$= \frac{i}{2\pi}\iint \frac{\exp[is(x - y)]}{\omega - ps/m + i\eta} M_{\omega}(y,p)dyds \tag{7}$$

$$M_{\omega}(y,p) = (2\pi)^{-1}\int ds\, e^{isy}\,\langle p + s\hbar/2| - i \quad (\omega + i\eta)\,\wp^{\text{in}}_{\omega}\,|p - s\hbar/2\rangle \tag{8}$$

Carrying out the s-integral in Equation 7 by contour integration and taking the $\eta \rightarrow 0$ limit, the scattered Wigner function for the stationary case, $\omega = 0$, has the following asymptotic limits:

$$\lim_{x\to\infty} W^{sc}_{0}(x,p)_{p>0} = -\lim_{x\to-\infty} W^{sc}_{0}(x,p)_{p<0} = \frac{m}{p}\int M_0(y,p)dy$$
$$\lim_{x\to-\infty} W^{sc}_{0}(x,p)_{p>0} = \lim_{x\to\infty} W^{sc}_{0}(x,p)_{p<0} = 0 \tag{9}$$

Using these relations and expressing the matrix elements of the transition superoperator in terms of matrix elements of the transition operator, the contribution to the scattered flux at large positive x for a pure initial state with definite "*positive*" momentum p′, $\wp^{\text{in}}_{0} = |p'\rangle\langle p'|$ takes the form:

$$j^{sc}_{x\to\infty} = -\frac{2\pi m}{p'\hbar}|T_{-p'p'}|^2 \tag{10}$$

To calculate the total flux at large positive x it is necessary to add the contribution from the incoming component of ρ:

$$j^{in} = \int_0^{\infty} \frac{p}{m} W^{in}_{p'p'}(x,p)dp = \frac{p'}{2\pi m\hbar} \tag{11}$$

where

$$W^{in}_{p'p'}(x,p) = \frac{1}{2\pi} \int e^{isx} \langle p + s\hbar/2|p'\rangle\langle p'|p - s\hbar/2\rangle \, ds$$
$$= \frac{1}{h} \delta(p - p') \qquad (12)$$

In summary, the transmitted flux j^T is given by:

$$j^T = -\frac{2\pi m}{p'\hbar}|T_{-p'p'}|^2 + \frac{p'}{2\pi m\hbar} \qquad (13)$$

and the scattered flux at large negative x, or reflected flux, j^R, is calculated by similar means:

$$j^R = j^{sc}_{x\to-\infty} = \int_{-\infty}^{0} dp \, \langle\langle \, p,p|i\mathcal{T}|p',p' \, \rangle\rangle = -\frac{2\pi m}{p'\hbar}|T_{-p'p'}|^2 \qquad (14)$$

The scattering (W^{sc}) contributions ot the flux at large positive and negative x are equal in agreement with flux conservation. Note moreover that $j^T = j^{in} + j^R$.

The reflectance coefficient $R_{p'}$ is the ratio between the magnitude of the reflected flux and the ingoing flux, while the transmission coefficient $T_{p'}$ is the ratio between the transmitted and incident flux. It follows from the above definitions that $j^T = j^{in}(1 - R_{p'})$ and $T_{p'} = 1 - R_{p'}$ as it should for total flux conservation.

Let us now examine the time-dependent case. The transmittance $\langle T\rangle$ is defined as the integral over all time of contributions of positive flux at asymptotically large x. (In Appendix B alternative expressions are provided.) The exponential factor $e^{-i\omega t}$ in Equation 2 and the integration over time in the definition of the transmittance provide a delta function $\delta(\omega)$ that precludes any of the nondiagonalities $\omega \neq 0$ from contributing:

$$\langle T\rangle = \lim_{x\to\infty} \int_{-\infty}^{\infty} dt \int_0^{\infty} \frac{p}{m} W(x,p,t)dp$$
$$= \lim_{x\to\infty} \int_{-\infty}^{\infty} dt \int_0^{\infty} dp\frac{p}{m} \iint W_{p'p''}(x,p)e^{-i\omega_{p'p''}t} \wp^{in}_{p'p''}(0)dp'dp''$$
$$= 2\pi \lim_{x\to\infty} \int_0^{\infty} dp\frac{p}{m} \iint W_{p'p''}(x,p)\delta(\omega_{p'p''}) \wp^{in}_{p'p''}(0)dp'dp'' \qquad (15)$$

It is convenient to divide this integral into incoming and scattered components, $\langle T\rangle = \langle T^{in}\rangle + \langle T^{sc}\rangle$, according to the decomposition $W_{p'p''}(x,p) = W^{in}_{p'p''}(x,p) + W^{sc}_{p'p''}(x,p)$. After some algebra one obtains for the free part:

$$\langle T^{in}\rangle = \int_0^\infty \wp^{in}_{p'p''}(0)dp' \tag{16}$$

In other words, in the absence of interaction, the transmittance is due to the positive part of the incoming momentum distribution. The scattered component becomes:

$$\begin{aligned}\langle T^{sc}\rangle &= 2\pi \int_0^\infty dp \int_{-\infty}^\infty \int_{-\infty}^\infty \langle\langle p,p| - i\mathcal{T}|p',p''\rangle\rangle\, \wp^{in}_{p'p''}(0)\delta(\omega_{p'p''})dp'dp'' \\ &= -2\pi i\hbar \int_0^\infty dp \int_0^\infty dp' \frac{m}{p'} [\langle\langle p,p|\mathcal{T}|p',p'\rangle\rangle\, \wp^{in}_{p'p''}(0) \\ &\quad + \langle\langle p,p|\mathcal{T}|-p',p'\rangle\rangle\, \wp^{in}_{-p'p'}(0) + \langle\langle p,p|\mathcal{T}|-p',-p'\rangle\rangle\, \wp^{in}_{-p'-p'}(0) \\ &\quad + \langle\langle p,p|\mathcal{T}|p',-p'\rangle\rangle\, \wp^{in}_{p'-p'}(0)]\end{aligned} \tag{17}$$

and the full transmittance is the sum of Equations 16 and 17. However, if the initial momentum distribution has a negligible negative momentum part:

$$\langle T\rangle = \int_0^\infty T_{p'} \wp^{in}_{p'p'}(0)dp' \tag{18}$$

i.e., the transmittance is just an average over the stationary transmission coefficients. Under these suitable conditions, the shape of the incoming state, pure or mixed, in coordinate space does not affect the transmittance (provided that the initial momentum distribution remains the same for the different shapes). This is in agreement with classical reasoning and numerical evidence[1] discussed in the next section. However, if some of the packet is directed away from the scattering center, there are interference terms contributing to the transmittance. These are on-the-energy-shell terms including directional coherences. Next we shall discuss this case in more detail.[8]

The formula [(Equation 16) + (Equation 17)] for transmittance $\langle T\rangle$ has been written in terms of the superoperator $\mathcal{T}$. In a more familiar language:

$$\begin{aligned}\langle T\rangle &= \int_0^\infty dp\langle p|S\wp^{in}(0)S^\dagger|p\rangle \\ &= \int_0^\infty dp\langle p|\wp^{out}|p\rangle = \int_0^\infty \langle p^{(-)}|\rho(0)|p^{(-)}\rangle\, dp\end{aligned} \tag{19}$$

$S = \Omega^{(-)\dagger}\Omega^{(+)}$ being the scattering operator, and $\wp^{out}$ the outgoing free density operator to which the full density operator ρ tends after the collision, $\Omega^{(-)}$ and $\Omega^{(+)}$ are the wave operators and $|p^{(-)}\rangle = \Omega^{(-)}|p\rangle$. Since both $\wp^{in}$

and $\wp^{out}$ are free-particle states, their diagonal in momentum matrix elements $\wp^{in}_{pp}$ and $\wp^{out}_{pp}$ are independent of time. That Equation 19 and our previous result are equal can be demonstrated by writing the matrix elements of $\mathcal{T}$ and of S in terms of matrix elements of the transition operator $T = V\Omega^{(+)}$.

In a real physical situation a density operator $\rho(0)$ is created at some instant of time (taken here as time 0) by some preparative procedure. For instance, one can create a packet in a higher potential surface by laser excitation from a lower one. The time $t = 0$ corresponds to the moment when the laser ceases pumping. The question then to be considered is what transmittance is to be associated with $\rho(0)$. It is tempting to equate $\rho(0)$ with $\wp^{in}(0)$ and apply Equation 19. However, $\wp^{in}$ is an ideal state that in the infinite past tends to the full ρ but evolves without the potential action. In order to make the identification between these two operators at time $t = 0$ they should be equal at all times $t < 0$; in other words, the real packet ρ should not feel the potential at $t < 0$. Moreover, even if the state $\rho(0)$ is located (in coordinate representation) asymptotically far from the potential center, this location does not automatically make it a *free* state in the scattering theory sense that would allow $\rho(0)$ and $\wp^{in}(0)$ to be equated. In particular, if $\rho(0)$ has a negative momentum component, this implies that in the past the packet was not free since the full ρ had collided in the past with the potential. (In general, when one refers to the "*past*", it must be understood as *the past that would occur if the present (time t = 0) Hamiltonian were the only one to consider at all times.)* Summarizing, a packet at an asymptotic position on the left with nonnegligible negative and positive momentum components is in a sense in the midst of a collision, and is to be described by the full density operator $\rho(0)$. Specifically, it cannot be equated to the free state $\wp^{in}(0)$.

Only for the special case that $\rho(0)$ has a negligible negative momentum component (and is asymptotically far to the left of the scattering center) $\rho(0)$ and $\wp^{in}(0)$ can be equated, and the expression for the transmittance reduces to

$$\langle T\rangle \approx \int_0^{\infty} dp T_p \rho(0)_{pp} \tag{20}$$

The above integral is a useful expression because there are efficient methods[2] to calculate the coefficients T_p. It is also the formula one would assume[1] based on classical intuition. However, in general, the first form of Equation 19 is not useful in practice since what we actually have is $\rho(0)$ and not $\wp^{in}(0)$. Therefore, the last integral in Equation 19 has to be used. This equation considers in principle diagonal and nondiagonal elements of $\rho(0)$, as well as positive and negative momentum contributions. The question then is whether the simpler formula in Equation 20 is a good approximation to the transmittance and in what circumstances. Clearly, if the initial negative momentum components of $\rho(0)$ are negligible (and $\rho(0)$ is asymptotically far from the scattering center), Equation 20 becomes exact. In fact, Equation 20 is generally valid provided that the packet $\rho(0)$ is initially far from the potential. To show this, $\langle p^{(-)}|\rho(0)|p^{(-)}\rangle$ in Equation 19 is decomposed into three terms having zero, one, and two matrix elements of the transition operator $T^{(-)} = V\Omega^{(-)}$ by using the Lippmann-Schwinger equation $\Omega^{(-)}|p\rangle = |p\rangle + (E_p - i0 - H_0)^{-1}T^{(-)}|p\rangle$. Inserting resolutions of the identity in coordinate representation in the second and third terms, assuming that the potential is far from the packet initial location, considering that T decays essentially as V does in coordinate space, and using the coordinate space representation of the resolvents, the one-dimensional optical theorem and the relation $T^{(-)}_{pp} = T^{*}_{pp}$ one obtains Equation 20.

III. CLASSICAL TRANSMITTANCE FORMULA

While the stationary transmission coefficients calculated for a definite incident momentum generally are very different in classical and quantum mechanics, the transmittances for time-dependent wave packets and the corresponding classical ensembles can be surprisingly close to each other due to the averaging effect made explicit in the previous section. This agreement frequently holds also in the tunneling region where the average energy $\langle E\rangle$ (quantum expectation value of the energy) is below the classical threshold (maximum energy of the barrier, V_0). It is worth stressing that such agreement is based on a statistical association of the wave packet with a classical ensemble, and that it would be lost if one adopted a one-particle interpretation of the classical limit of the packet. The present treatment follows Muga[1] and assumes an initial Gaussian packet with initial average momentum p_0.

A systematic study of the relation between classical statistical and quantum transmittance is conveniently carried out by introducing reduced dimensionless quantities. Only the two dimensionless parameters f and r determine the behavior of the system classically and to a large extent also quantum mechanically:

$$r = \frac{\langle E\rangle}{V_0}, \qquad f = \frac{p_0^2}{2m\langle E\rangle} \tag{21}$$

m being the mass of the particle. Also of importance, especially in the quantum case, is the dimensionless potential strength $U_0 = (2mV_0)/(\hbar^2 a^2)$, where a^{-1} is the characteristic length (width) of the barrier. A classical transmittance formula is obtained by associating the initial Gaussian with a classical ensemble with the same momentum distribution. Then, the transmission probability is the fraction of the ensemble with energy above the classical threshold:[1]

$$T = 1 - \mathcal{P}\,[(1 - M)/S] \tag{22}$$

where $\mathcal{P}[(1 - M)/S]$ is the probability integral of the Gaussian distribution $\mathcal{Z}[(\Pi - M)/S]$ having mean $M = (fr)^{1/2}$ and variance $S^2 = (1 - f)r$. Π is the dimensionless momentum $\Pi = p/(2mV_0)^{1/2}$.

A number of consequences follow from this simple expression:

1. The transmittance does not depend on the initial position of the packet, provided it is far enough from the potential and the rest of the Gaussian parameters are fixed.
2. The transmittance does not depend on the strength parameter U_0 or the shape of the potential.
3. The transmittance depends only on the dimensionless parameters f and r.
4. The transmittance surface has a maximum at $f = r$, for $f < 1$.

The validity of the quantum parallel of the first item is justified by the discussion in the previous section. If the transmittance depends only on the initial momentum distribution (Equation 20), then no matter where the packet is located initially (within the left asymptotic side) the initial momentum distribution and therefore the transmittance do not change. Numerical propagations of wave packets have been performed to check the validity of these "predictions" in the quantum case. (The reader is referred to the original article by Muga[1] for details on the comparison.) Here it is enough to conclude that these four points are fulfilled with remarkable quantitative accuracy by the quantum dynamics except in the following cases: in the stationary limit (f close to 1), for low U_0, for discontinuous potentials, and for resonance widths (if resonances are present) that are larger than the packet width. The purely classical results are even better than semiclassical methods based on the assumption that the packet evolves preserving a Gaussian form. The simplest, "local" version considers a classical motion for the average position and momentum. Thus tunneling is impossible. A modified method making use of "global" properties of the state to determine the evolution of the Gaussian moments overcomes part of this difficulty.[9]

Note that for obtaining the transmittance formula in Equation 22 no assumption has been made on the initial classical ensemble other than the agreement with the initial quantum marginal distribution for momentum. The classical ensemble can be further specified by imposing agreement with the initial quantum Wigner function (classical Wigner method). This idea has been exploited for a series of applications in molecular physics.[11–16] To specify the full distribution is useful for obtaining the full evolution, and not only the transmittance, which is an asymptotic property. We have implemented the classical Wigner method to find whether the evolution of moments of the classical and quantum distribution are close to each other. In Figures 1 to 4, the different moments are displayed against time for a $\cosh^{-2}(x)$ potential with collision parameters $r = 1$, $f = 0.901$, $U_0 = 10$. (This corresponds to a point on the left edge of Figure 1 in Muga.[1] A *single* classical particle with the average energy of the packet would never cross the barrier.) These have

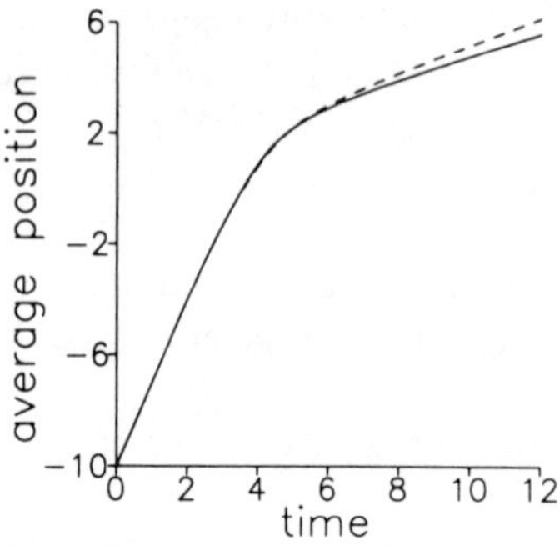

FIGURE 1. Classical (solid) and quantum (dashed) average position vs. time. ($m = 1$, $V_0 = 5$, atomic units.)

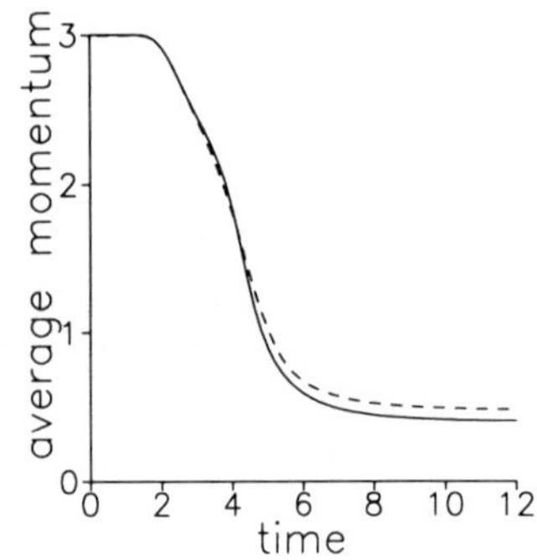

FIGURE 2. Classical (solid) and quantum (dashed) average momentum vs. time. ($m = 1$, $V_0 = 5$, atomic units.)

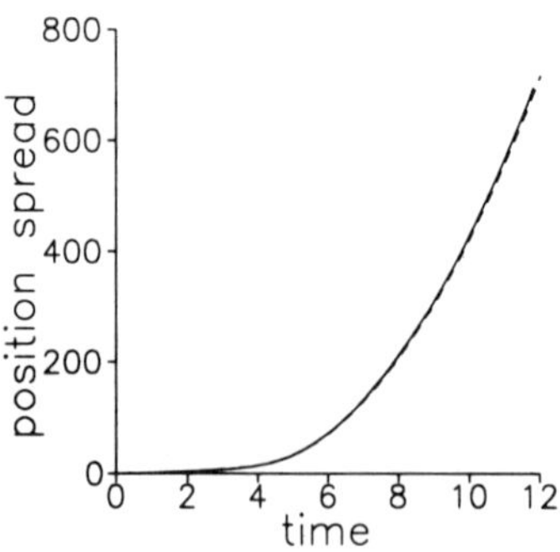

FIGURE 3. Classical (solid) and quantum (dashed) position spread (variance) vs. time. ($m = 1$, $V_0 = 5$, atomic units.)

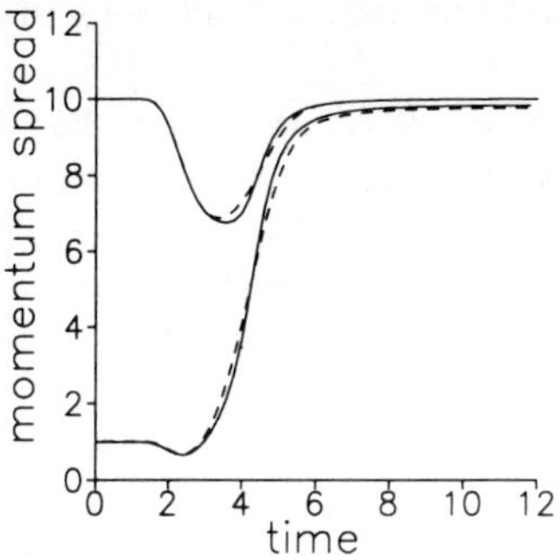

FIGURE 4. Above: Classical (solid) and quantum (dashed) momentum spread (second moment about the origin). This is proportional to the average kinetic energy. Below: Classical (solid) and quantum (dashed) variance (central second moment). ($m = 1$, $V_0 = 5$, atomic units.)

been chosen so that the transmittances are close to each other (.45 and .44). The figures show that the agreement is not only asymptotic but extends to details of the time-dependent propagation.

While the above statistical treatment is approximate, exact results can be achieved by means of classical trajectories. The work of Papadopoulos[3] is based on writing the exact quantum propagator as a product of the semiclassical Van Vleck propagator K_c (which is a function of the classical action and its second order partial derivative) times a correction exponential factor:

$$K(xt|x'0) = K_c(xt|x'0)\exp[iQ(xt|x'o)/\hbar] \tag{23}$$

$$Q = \sum_{n=2}^{\infty} \hbar^n Q_n \tag{24}$$

The partial phases $Q_n(xt|x'0)$ are obtained from the classical path $X_c(xt|x'0; \tau)$ satisfying the end conditions $X_c(0) = x'$ and $X_c(t) = x$. The procedure for doing so involves a hierarchy of equations. The method is an interesting example of the possibilities of classical structures in solving quantum problems.

IV. QUANTUM STREAMLINES

It is possible to generalize the classical concept of trajectory in quantum mechanics for attaining all the relevant quantum stationary scatttering information in the one-dimensional systems we are studying, in particular, the stationary transmission coefficients in the tunneling regime, phase shifts, dwell times, etc. Actually, the ''quantum trajectories'' in the formalism to be presented are physically hydrodynamic streamlines rather than real particle paths, in spite of the fact that they are solutions of modified phase space Hamiltonian equations.

In this section we shall sketch the main ideas behind the streamline method:[2] the approach is very simple to implement, since it is an initial value problem, and accurate, as tested with several potential models.

The time-dependent Schrödinger equation in the coordinate representation can be split into two coupled real equations by means of the substitution:

$$\psi = \exp(iW/\hbar); \qquad R = \Re e W,\ I = \Im m W \tag{25}$$

for the wave function, and separation of real and imaginary parts:

$$\frac{\partial R}{\partial t} + \frac{1}{2m}[(R^{(1)})^2 - (I^{(1)})^2 + \hbar I^{(2)}] + V = 0 \tag{26}$$

$$\frac{\partial I}{\partial t} + \frac{R^{(1)}I^{(1)}}{m} - \frac{\hbar R^{(2)}}{2m} = 0 \tag{27}$$

The superscripts in parentheses denote the order of the derivative with respect to the position. The first equation is usually interpreted as a Hamilton-Jacobi-like equation, where the potential is given by the "classical term" V plus a "quantum potential":

$$Q(x,t) = \frac{1}{2m}[\hbar I^{(2)} - (I^{(1)})^2] \tag{28}$$

An alternative interpretation regards Equation 26 as the hydrodynamic equation of motion in terms of a velocity potential.[17] (This has the advantage with respect to the usual Hamilton-Jacobi interpretation that the mysterious quantum potential has a classical analog, namely, the internal hydrodynamic potential due to the nonzero width of the momentum distribution.[17]) The second equation is a continuity equation. The pair of Equations 26 and 27 is taken as the starting point for constructing a dynamic hierarchy of equations equivalent to the Schrödinger equation.

The Hamilton-Jacobi equation in coordinate space is equivalent to Hamilton's equation in phase space:

$$\dot{q} = \frac{R^{(1)}}{m} \tag{29}$$

$$\dot{R}^{(1)} = -V^{(1)} - Q^{(1)} \tag{30}$$

Here $q = q(t)$ is the path followed by the trajectory in coordinate space. In Hamilton's equations $R^{(1)}$ is an independent variable on the same footing as q, except for the fact that the initial conditions are restricted according to $q(0) = x_0$ and $R^{(1)}(0) = R^{(1)}(x_0,0)$. The same comment applies later to $I^{(n)}$ and $R^{(n)}$.

Our objective now is to solve Hamilton's equations (Equations 20 and 30) for given initial conditions, as it is done in standard classical mechanics. However, the quantum potential is not known *a priori*. (Q depends on the solution through $I^{(1)}$ and $I^{(2)}$.) A way out is to obtain differential equations for the variables $I^{(1)}$ and $I^{(2)}$ entering in the definition of Q by means of total derivatives with respect to time along the q(t) path (substantial derivatives). For a field variable f(q,t) the substantial derivative is

$$\dot{f} = \frac{\partial f}{\partial t} + \frac{\partial f}{\partial q}\frac{dq}{dt} \tag{31}$$

Using Equation 31, the derivatives of Equation 27 with respect to the position and Leibniz's theorem, one obtains the general expression:

$$\dot{I}^{(n)} = \frac{\hbar R^{(2+n)}}{2m} - \frac{1}{m}\sum_{r=1}^{n}\binom{n}{r} I^{(n-r+1)}R^{(r+1)}, \quad n \geqslant 1 \tag{32}$$

Similarly:

$$\begin{aligned}\dot{R}^{(n+1)} &= -V^{(n+1)} - Q^{(n+1)} \\ &\quad - \frac{1}{2m}\sum_{r=1}^{n}\binom{n+1}{r} R^{(n+2-r)}R^{(r+1)}, \quad n \geqslant 1\end{aligned} \tag{33}$$

where the nth derivative of the quantum potential is given by:

$$Q^{(n)} = \frac{\hbar I^{n+2)}}{2m} - \frac{1}{m}\sum_{r=0}^{n-1}\binom{n-1}{r} I^{(n-r)}I^{(r+2)}, \quad n \geqslant 1 \tag{34}$$

The set of Equations 29, 30, 32, 33, and the equations:

$$\dot{R} = \frac{1}{2m}[(R^{(1)})^2 + (I^{(1)})^2 - \hbar I^{(2)}] - V \tag{35}$$

$$\dot{I} = \frac{\hbar R^{(2)}}{2m} \tag{36}$$

which follow from the substantial derivation of Equations 26 and 27, form an infinite hierarchy of first order differential equations equivalent to the Schrödinger equation: the solution of the system runs in the infinite dimensional space $\{q, R^{(n)}, I^{(n)}; n = 0, 1, 2, \ldots\}$, and its projection onto the subspace {q, R, I} yields the exact value of the wave function at q(t).

An infinite hierarchy is impossible to solve unless some truncation is done. Imposing the condition $I^{(n)} = R^{(n)} = 0$, $n > n_0$ for larger and larger n_0 has not led to converged results for time-dependent collisions, and at this point it is not clear how the truncation should be performed in the general time-dependent case. There are nevertheless obvious choices for particular potentials. Thus, for harmonic or linear interactions, the Gaussian form of the packet is preserved and derivatives of third order or higher of R and I vanish. As a consequence a finite set of equations gives the exact quantum propagation if the second derivatives are included in the finite set of variables.

Another important case where the hierarchy can be exactly closed is stationary scattering: if the wave function is stationary, the energy of the trajectory is conserved. This conservation provides a way to check the accuracy of the numerical integration:

$$\frac{(R^{(1)})^2}{2m} + V + Q = E \tag{37}$$

The relevant substantial derivatives take the form:

$$\dot{R} = \frac{(R^{(1)})^2}{m} - E; \quad \dot{I} = \frac{I^{(1)}R^{(1)}}{m} \tag{38}$$

$$\dot{R}^{(1)} = \frac{2}{\hbar m} I^{(1)}(R^{(1)})^2 \tag{39}$$

$$\dot{I}^{(1)} = \frac{R^{(1)}}{m\hbar}[2mE + (I^{(1)})^2 - (R^{(1)})^2 - 2mV] \tag{40}$$

Equations 29, 38, 39, and 40 form a closed set of five differential equations that determine the exact wave function along the path q(t). (The number of coupled equations can be reduced to four by a mathematically related but less physically appealing method described in the appendix of Muga.[1]) If one is interested only in the transmission coefficient, the equation for R is not needed. It is enough to calculate one trajectory per energy to determine entirely the wave function becuase all the possible trajectories in phase space have to run along the curve $p = R^{(1)}(x)$; since the motion in phase space is governed by Hamiltonian dynamics, two moving phase points passing over the same point in phase space follow the same trajectory (possibly at different times).

The method is suitable for stationary scattering states of the continuum, being invalid for ''static'' bound states where all the substantial derivatives vanish. Another restriction is given by the requirement of continuous potentials with continuous first derivatives (see Equations 30 and 40).

With respect to the choice of initial conditions, consider the asymptotic form of the continuum eigenstate of H with incident positive momentum. The intial conditions cannot be fixed on the left side of the barrier since we ignore *a priori* the complex constants of the asymptotic solution. However, on the right the form of the solution is exactly known once a unit outgoing density normalization is adopted. Therefore, the initial conditions are set in this region and the trajectories run backward, with negative momentum. (This amounts simply to working with ψ^* rather than with ψ.)

The transmission coefficient is then calculated by averaging $\psi\psi^* = \exp(-2I/\hbar)$ asymptotically as $x \rightarrow -\infty$, i.e., eliminating the oscillatory terms.

Checks have been done comparing analytically known transmittances with the results obtained following this procedure. We found excellent agreement (see Figure 1 of Reference 2).

From a practical point of view, the advantage of the method rests on the fact that solving the Schrödinger equation becomes an initial value problem, which is very easy to implement by using standard subroutines for coupled first order differential equations. Other methods are more inconvenient or inaccurate. Thus the WKB method is just an approximation, while discrete methods are generally harder to work with.

The flux along trajectories in coordinate space is just the quantum current density (see Equation 42 below), so they can actually be regarded as streamlines. We shall show that the time spent by the trajectory between two points is related to the dwell time, a measure of the average time spent by the particle in a region of space:

$$\tau_D(a,b) = \frac{1}{j^{in}} \int_a^b |\psi(x)|^2 dx \tag{41}$$

where j^{in} is the ingoing flux. The flux (current density) is given by:

$$J = \frac{|\psi|^2 R^{(1)}}{m} = \frac{\hbar}{m} \Im m[\psi^* \partial\psi/\partial x] \tag{42}$$

This quantity is constant along a trajectory, as can be seen from Equations 38 and 39, and equals the outgoing transmitted flux j^T. Integration over time for a given trajectory leads to:

$$\int_{t_1}^{t_2} J(t)dt = j^T t_{12}; \quad t_{12} = t_2 - t_1 \tag{43}$$

However, the integral can also be written as:

$$\int_{t_1}^{t_2} J(t)dt = \int_{a=q(t_1)}^{b=q(t_2)} |\psi(q)|^2 dq = \tau_D(a,b) j^{in} \tag{44}$$

by using the expression in Equation 42 for $J = j^T$, the formal change of variable $q = q(t)$ and Equation 41. Comparison of Equations 43 and 44 leads to $t_{12} = \tau_D(a,b)/T$, i.e., the trajectory time is proportional to the dwell time; the proportionality factory is the ratio between ingoing and transmitted flux, which is by definition the inverse of the transmittance. This is an interesting result because it gives physical content to the trajectories and provides a simple and economical way for calculating dwell times. Observe that calculating t_{12}, just the difference between two times, Equation 43, is faster and more accurate than obtaining τ_D from its standard definition as an integral over the coordinate of the density, Equation 41.

As an application we have evaluated the dwell time and the excess dwell time, $\tau_E = \tau_D - (a - b)m/(k\hbar)$, as a function of the incident energy between the maxima of a double barrier (Figure 2, Reference 2). A resonance is clearly identified in this case, both from the dwell times and from the transmittance peak. Since the transmittance peaks do not always agree with dwell time maxima,[18] it is important, in general, to evaluate at least both quantities to identify unambiguously a resonance.

For a complementary recent discussion on one dimensional scattering in terms of the invariants of quantum hydrodynamics see Reference 19.

There is another purely classical approach worth commenting on which is qualitatively different from other semiclassical methods: it is called classical high-jumping[20] and is based on assigning a nonvanishing length to the classical bodies (“ropes”) of the ensemble. For a finite length rope it is classically possible to overcome a potential barrier of height V_0 with initial kinetic energies below V_0. To avoid a step-like transmittance in the stationary case, it has been proposed that rather than a fixed length there should exist a distribution of lengths. The theory could be improved if some kind of *a priori* justification of the length distribution were found. Another problem to think about is that at energies higher than V_0 *all* the ropes pass the barrier, no matter their length. Thus, even if the tunneling is somehow accounted for, at least for nonresonant potentials, the regions above the barrier energy is not properly described. A way out is to add internal structure to the ropes. However, this brings new questions without obvious answers. What is the internal potential like and why?

V. TUNNELING TIMES

In spite of a great deal of effort to define and measure tunneling times, the question of how long it takes to tunnel through a barrier seems to elude a definitive answer. Different times have been proposed, and a recent critical review discusses *and rejects* the main contenders.[21] A basic criterion that has been suggested[21] for accepting a tunneling time theory is the separability of the dwell time (here defined for time-dependent wave packets with probability density $\varrho(x)$; the density operator ρ and ϱ should not be confused:

$$\tau_D(a,b) = \int_{-\infty}^{\infty} dt \int_a^b \varrho(x)dx \tag{45}$$

into transmitted and reflected components according to:

$$\tau_D = \tau_T T + \tau_R R \tag{46}$$

where τ_T and τ_R are, respectively, the average time spent in the region $a < x < b$ by transmitted and reflected particles, and $R = 1 - T$ is the reflectance.

In this section we shall outline a possible solution to this problem. An extensive discussion will be given elsewhere.[22] The basic idea is that the formal answer given by classical mechanics to the above question can be translated into quantum mechanics with a reinterpretation of the symbols involved.

Suppose that a classical ensemble of particles characterized by a phase space distribution function $f = f(p,q,t)$ (normalized to one) collides with a potential barrier. The average time spent by a particle between $q = a$ and $q = b$ can be written in several ways (the integral limits go from $-\infty$ to ∞ if it is not indicated otherwise):

$$\tau_D = \iint f(p_0,q_0,0)\tau(a,b,p_0,q_0)dp_0dq_0$$

$$= \int dq_0 \int dp_0 \int dt\ f(p_0,q_0,0)\{\mathcal{H}[q(p_0,q_0,t) - a] \tag{47}$$

$$-\mathcal{H}[q(p_0,q_0,t) - b]\} \tag{48}$$

$$= \int dt \int dq \int dp\ f(p,q,t)[\mathcal{H}(q - a) - \mathcal{H}(q - b)] \tag{49}$$

$$= \int dt \int_a^b dq\ \varrho(q) \tag{50}$$

where $\tau(a,b,p_0,q_0) = \int\{\mathcal{H}[q(q_0,p_0,t) - a] - \mathcal{H}[q(q_0,p_0,t) - b]\}dt$ is the time spent between a and b by the trajectory with initial conditions (momentum and position) (p_0,q_0), and $\mathcal{H}$ is the Heaviside function. The expressions in Equations 47 and 48 make use of the phase space of initial conditions. They correspond to a Heisenberg-like picture where the dynamics are taken into account by the observable (note the time dependence in the Heaviside functions). The last two equations are written in terms of the space of moving phase points (Schrödinger-like picture), and the explicit time dependence has been translated to the distribution function. The transformation between p_0,q_0, and p,q is canonical so $dqdp = dq_0dp_0$. Besides, Liouville's theorem states that $f(p,q,t) = f(p_0,q_0,0)$, where (p,q) is the phase at time t of the trajectory starting at (p_0,q_0) at time $t = 0$. In Equation 50 $\varrho(q)$ is the integral over momentum of the distribution f(q,p), i.e., the local probability density.

Clearly, Equations 45 and 50 have the same form. Actually, the set of Equations 47 to 50 are meaningful in quantum mechanics by taking f(p,q) as the Wigner function. The preceding paragraph is equally valid by substituting "trajectory" by "Wigner trajectory". A Wigner trajectory is a generalization of the classical trajectory.[14,23] Along a Wigner trajectory the Wigner function keeps a constant value.

The initial phases can be divided into two sets: those leading to a transmitted (classical or Wigner) trajectory, and those leading to a reflected one. This separates the phase space plane into two domains, D_T and D_R, respectively, and the distribution function $f(p_0,q_0,0)$ into a part that will pass the barrier, $f_T = fP_T$, and a part to be reflected, $f_R = (1 - P_T)f$, where P_T selects the points in the D_T domain:

$$
\begin{aligned}
P_T(p_0,q_0) &= 1 \text{ if} (p_0,q_0) \in D_T \\
P_T(p_0,q_0) &= 0 \text{ otherwise}
\end{aligned}
\tag{51}
$$

These partial distributions for transmitted and reflected trajectories can be normalized to one by dividing by T and R, respectively, The average time spent by the transmitted trajectories between a and b becomes $\tau_T = T^{-1} \int dp_0 \int dq_0 P_T f(q_0,p_0,0)\tau(a,b,p_0,q_0)$, with a similar expression for the average time for reflected trajectories τ_R. Equation 46 then follows immediately from Equation 47 by separating the double integral into the two domains and using the expressions for τ_T and τ_R. τ_T may be regarded as dwell time for the normalized part of the packet that passes the barrier, with a similar interpretation for τ_R. One, however, should be cautious in the physical interpretation of the formal results. Wigner trajectories are yet quite unexplored, and further work is needed to clarify the way in which these structures can be used for quantum mechanical computations.

Recently Jensen and Buot[24] have used the Wigner trajectories for evaluating tunneling times in stationary scattering. The treatment given above is for time-dependent tunneling, and demonstrates the separation of the dwell time into reflected and transmitted parts.

VI. DISCUSSION AND CONCLUSIONS

Classical mechanics is an inspiring source of methods, structures, and concepts for quantum mechanics. It is not by chance that different formulations of quantum mechanics have to a great extent a classical parallel: Feynman's path integral approach is connected with Lagrangian mechanics, while Bohm's causal interpretation follows Hamilton-Jacobi theory; the standard formulation is indebted to the Hamiltonian theory and to the theory of canonical transformations; and the Weyl-Wigner phase space equivalent representation is associated with classical statistical mechanics.[25–28] In other words, classical mechanics is present in the quantum theory not only as a limit or in the form of semiclassical approximations. It is indeed part of the skeleton of quantum mechanics. In this regard, a phenomenon as quantum as tunneling is not an exception and admits a classical perspective.

This review summarizes a series of works that have in common certain classical points of view of the quantum tunnel effect. We have purposely set aside "semiclassical methods" (WKB) to concentrate on either purely classical approximations to the quantum results or exact quantum approaches based on classical structures. In summary:

1. Scattering in one dimension has been formulated in phase space for stationary and time-dependent conditions. Special attention has been paid to the difficulties when the initial packet has negative momentum components. A general formula has been provided for the transmittance, and the domain of applicability of the simple approximate expression one expects on classical grounds (the average of stationary transmission coefficients) has been indicated.
2. A classical statistical approach has been shown to give excellent results for the transmittance even below the classical threshold. The cases where this is not so have been explained.
3. All the scattering information, including the exact wave function, can be obtained by running a classical-like trajectory in an extended space for stationary states. In the time-dependent case, purely classical trajectories area also under a suitable treatment able to provide the exact quantum propagation.
4. Wigner trajectories allow the separation of the dwell time into real transmitted and reflected contributions, the formalism being common for quantum and classical mechanics.*

Our group is currently working to extend the above results in various directions. Applications to electronic devices include the numerical evaluation of tunneling times and inclusion of relativistic effects.

* We have recently discovered that Liouville's theorem does not hold in general for Wigner trajectories. Alternative decompositions of the dwell time are discussed in Muga, J. G., Brouard, S., and Sala, R., *J. Phys. Condens. Matter,* 4, L579, 1992.

ACKNOWLEDGMENTS

We acknowledge R. F. Snider for his discussions and encouragement. The work has been partly supported by a grant from Universidad de La Laguna.

APPENDIX A: SCATTERING THEORY RELATIONS

First operator relations and then superoperator relations are discussed. The abstract Möller and transition operators are defined as:

$$\Omega = \lim_{t \to -\infty} e^{iHt/\hbar} e^{-iH_0t/\hbar}, \quad T = V\Omega \tag{A.1}$$

Another frequent notation for Ω is $\Omega^{(+)}$ to distinguish it from $\Omega^{(-)}$. ($\Omega^{(-)}$ is defined as Ω is but with $+\infty$ instead of $-\infty$ in the time limit.)

They are related to the parameterized operators by:

$$\Omega\phi = \int \lim_{\epsilon \to 0} \Omega(E_p + i\epsilon)|p\rangle \langle p|\phi\rangle dp \tag{A.2}$$

$$T|p\rangle = \lim_{\epsilon \to 0} T(E_p + i\epsilon)|p\rangle \tag{A.3}$$

where

$$\begin{aligned} \Omega(z) &= 1 + \frac{1}{z - H_0} T(z) \\ T(z) &= V + V\frac{1}{z - H} V \end{aligned} \tag{A.4}$$

The abstract Möller and transition superoperators are defined similarly as:

$$\begin{aligned} \Omega_L &= \lim_{t \to -\infty} e^{i\mathcal{L}t} e^{-i\mathcal{L}_0 t} \\ \mathcal{T} &= \mathcal{V}\Omega_L \end{aligned} \tag{A.5}$$

where the Liouville superoperator $\mathcal{L} = \hbar^{-1}[H, \ ]$ is decomposed into its free and potential parts, $\mathcal{L} = \mathcal{L}_0 + \mathcal{V}$:

$$\begin{aligned} \mathcal{L}_0 &= \hbar^{-1}[H_0, \ ] \\ \mathcal{V} &= \hbar^{-1}[V, \ ] \end{aligned} \tag{A.6}$$

The abstract Ω_L and $\mathcal{T}$ can also be expressed in terms of Ω and T as:

$$\begin{aligned} \Omega_L A &= \Omega A \Omega^\dagger \\ \hbar \mathcal{T} A &= T A \Omega^\dagger - \Omega A T^\dagger \end{aligned} \tag{A.7}$$

They are also related to the parameterized superoperators by:

$$\Omega_L A = \int \lim_{\eta \to 0} \Omega_L(\omega_{pp'} + i\eta)|p,p'\rangle\rangle\langle\langle pp'|A\rangle\rangle \, dp \, dp' \tag{A.8}$$

and

$$\mathcal{T}\,|p,p'\rangle\rangle = \lim_{\eta \to 0} \mathcal{T}(\omega_{pp'} + i\eta)|p,p'\rangle\rangle \tag{A.9}$$

where

$$\begin{aligned} \Omega_L(z) &= 1 + \frac{1}{z - \mathcal{L}_0} \mathcal{T}(z) \\ \mathcal{T}(z) &= \mathcal{V} + \mathcal{V} \frac{1}{z - \mathcal{L}} \mathcal{V} \end{aligned} \tag{A.10}$$

The double brackets used for tetradic matrices of superoperators are given in terms of the usual brackets by:

$$|p,p'\rangle\rangle = |p\rangle\langle p'| \tag{A.11}$$

$$\langle\langle p,p'| \ldots = \langle p| \ldots |p'\rangle \tag{A.12}$$

APPENDIX B: TRANSMITTANCE EXPRESSIONS

The transmittance in time-dependent scattering can be expressed in different ways. Here it is shown that a form suitable for numerical computations is equivalent to the definition given in Section I. Other expressions found by Turner and Snider[9] are also reported. It is assumed that the initial state is fully localized on the left side of the potential barrier.

A common definition for transmittance is *the probability to find the particle at* $x > y >> 0$ *at a time* τ *large enough so as to make this probability independent of time*. Thus $\rho(t)$ has become equal $\wp^{out}(t)$ for $t \geq \tau$. This may be written in different ways:

$$\begin{aligned}\langle T\rangle &= \int_y^{\infty} dx\rho(\tau)_{xx} = \int_{-\infty}^{\infty} dp \int_y^{\infty} dx W^{out}(x,p,\tau) \\ &= \int_0^{\infty} dt \int_{-\infty}^{\infty} dp \frac{p}{m} W^{out}(y,p,t)\end{aligned} \tag{B.1}$$

where the last equality follows from integrating the quantum continuity equation twice (once over position and once over time) with the understanding that $W^{out}(x,p,t)$ vanishes at $x = \infty$; moreover, at $t = 0$, $W^{out}(y,p,0)$ vanishes (or is negligibly small), since none of the packet has yet reached position y. Finally, the upper time integration limit has been shifted to ∞. The first integral in Equation B.1 is useful when the numerical propagation of the wave packet is feasible. Connection with the previous definition and Equation 19 is made by noticing that, due to the rapidly oscillating exponential factor in

$$W^{out}(x,p,t) = \frac{1}{2\pi}\int e^{is(x-pt/m)} < (p + s\hbar/2)^{(-)}|\rho(0)|(p - s\hbar/2)^{(-)} > ds$$

the lower integration limits for t and p in Equation B.1 may be changed to $-\infty$ and 0, respectively.

Two more more interesting expressions for the transmittance have been found by Turner and Snider;[9] they first show by superoperator techniques that:

$$\langle T\rangle = \int_0^{\infty} dpG(p,\infty) \tag{B.2}$$

where $G(p,\infty)$ is the momentum distribution at infinite time, i.e., $G(p,\infty) = \wp_{pp}^{out}$ in agreement with Equation 19. Then, assuming that the state is a pure one, one has $G(p) = |\phi(p)|^2$, and Parseval's equality permits writing the last integral as a position integral involving the inverse Fourier transforms of ϕ^* and $\mathcal{H}(p)\phi(p)$, where $\mathcal{H}$ is the Heaviside function. Since the last inverse transform is a convolution, the resulting expression is a double integral:

$$T = \frac{1}{2} + \frac{1}{2\pi}\int_{-\infty}^{\infty}\int_{-\infty}^{\infty} \frac{\mathrm{Im}\psi(y)^*\psi(x)}{x - y} dxdy \tag{B.3}$$

REFERENCES

1. **Muga, J. G.,** Classical transmittance and tunnelling, *J. Phys. A Gen. Phys.*, 24, 2003, 1991.
2. **Muga, J. G.,** Stationary scattering as an initial value problem, *Phys. Lett. A*, 157, 325, 1991; 159, 442, 1991(E).
3. **Papadopoulos, G. J.,** Time-dependent quantum tunnelling via crossover processes, *J. Phys. A.*, 23, 935, 1990.
4. **Muga, J. G. and Snider, R. F.,** Asymptotic behaviour in phase space scattering, *Phys. Rev. A*, 45, 2940, 1992.
5. **Carruthers, P. and Zachariasen, F.,** Quantum collision theory with phase space distributions, *Rev. Mod. Phys.*, 55. 245. 1983.
6. **Snider, R. F.,** Differential cross section from an operator viewpoint, *J. Chem. Phys.*, 63, 3256, 1975.
7. **Turner, R. E. and Snider, R. F.,** Superoperator methods of calculating cross-sections. III. Unified description of classical and quantum scattering, *Can. J. Phys.*, 58, 1171, 1980.
8. **Muga, J. G., Brouard, S., and Snider, R. F.,** Transmittance for wave-packet scattering, *Phys. Rev. A*, 46, 6075, 1992.
9. **Turner, R. E. and Snider, R. F.,** A comparison of local and global single Gaussian approximations to time dynamics. One dimensional systems, *J. Chem. Phys.*, 87, 910, 1987.
10. **Heller, E. J.,** Wigner phase space method: analysis for semiclassical applications, *J. Chem. Phys.*, 65, 1289, 1976.
11. **Lee, H. W. and Scully, M. O.,** A new approach to molecular collisions: statistical quasiclassical method, *J. Chem. Phys.*, 73, 2238, 1980.
12. **Brown, R. C. and Heller, E.,** Classical trajectory approach to photodissociation: the Wigner method, *J. Chem. Phys.*, 75, 186, 1981.
13. **Heller, E. and Brown, R. C.,** Errors in the Wigner approach to quantum dynamics, *J. Chem. Phys.*, 75, 1048, 1981.
14. **Lee, H. W. and Scully, M. O.,** Wigner phase-space description of a Morse oscillator, *J. Chem. Phys.*, 77, 4604, 1982.
15. **Davis, M. J. and Heller, E.,** Comparisons of classical and quantum dynamics for initially localized states, *J. Chem. Phys.*, 80, 5036, 1984.
16. **Lee, H. W. and George, F.,** Wigner phase-space description above and below the classical threshold for the $H + H_2$ reaction, *J. Chem. Phys.*, 84, 6247, 1986.
17. **Muga, J. G. and Snider, R. F.,** to appear.
18. **Muga, J. G. and Snider, R. F.,** Scattering by a separable potential in one dimension, *Can. J. Phys.*, 68, 403, 1990.
19. **Nassar, A. B.,** *Phys. Lett. A*, 146, 89, 1990.
20. **Cohn, A. and Rabinowitz, M.,** Classical tunneling, *Int. J. Theor. Phys.*, 29, 215, 1990.
21. **Hauge, E. H. and Stovneng, J. A.,** Tunneling times: a critical review, *Rev. Mod. Phys.*, 61, 917, 1989; **Büttiker, M.,** in *Electron Properties of Multilayers and Low-Dimensional Semiconductor Structures*, Chamberlain, J. M. et al., Eds., Plenum Press, New York, 297, 1990; **Landauer, R.,** *Ber. Bunsenges. Phys. Chem.*, 95, 404, 1991.
22. **Muga, J. G., Brouard, S., and Sala, R.,** Transmission and reflection tunneling times, *Phys. Lett. A*, 167, 24, 1992.
23. **Lee, H. W.,** Quantum potential in the Wigner phase space, *Phys. Lett. A*, 146, 287, 1990.
24. **Jensen, K. L. and Buot, F. A.,** Numerical calculation of particle trajectories and tunneling times for resonant tunneling barrier structures, *Appl. Phys. Lett.*, 55, 669, 1989.
25. **Takabayasi, T.,** On the formulation of quantum mechanics associated with classical pictures, *Prog. Theor. Phys.*, 8, 143, 1952.

26. **Takabayasi, T.,** Remarks on the formulation of quantum mechanics with classical pictures and relations between linear scalar fields and hydrodynamical fields, *Prog. Theor. Phys.*, 9, 187, 1953.
27. **Takabayasi, T.,** The formulation of quantum mechanics in terms of ensembles in phase space, *Prog. Theor. Phys.*, 11, 341, 1954.
28. **Hillery, M., O'Connell, R. F., Scully, M. O., and Wigner, E. P.,** Distribution functions in physics: fundamentals, *Phys. Rep.*, 106, 121, 1984.

Chapter 6

PHOTOELECTROCHEMICAL DEVICES UTILIZING MOLECULE-BASED EXCITED STATE REACTIONS AT COATED POLYMER MEMBRANES

Ru-Juang Lin and Masao Kaneko

TABLE OF CONTENTS

0-8493-8061-8/93/$0.00 + $.50

I. INTRODUCTION

Photoelectrochemical processes are attracting much attention not only in photochemical energy conversion but also in photodevices which are important for the future so-called photonics.[1-6] At present, photodevices utilize photophysical processes of semiconductors. They are based on continuous bulk properties of semiconductors, so that the minimum size of the device is determined by the minimum continuous structure of the semiconducting material to exhibit the objective photophysical properties. In order to design molecular size devices, it is important to utilize at least molecule-based processes. This is also the case when designing molecular photoelectrochemical devices. For this purpose, a photoexcitation center compound should be combined with other molecules to realize designed photofunctions based on molecules. Electron transfer at the excited state to the acceptor molecule or from the donor molecule is a typical process utilizable for photoelectrochemical devices.

In the studies utilizing molecule-based photochemical reactions for photoelectrochemical devices, it has been shown that simple homogeneous solution systems have few prospects of being applied to photodevices, because the rate of back electron transfer is usually too rapid to obtain net photoelectrochemical response. This problem can be overcome by using heterogeneous systems.[5-8] It has been demonstrated that immobilization of one or all the photochemical reaction components onto an electrode was effective in fabricating photodevices even if the solution photochemistry of the same compounds cannot lead to photoelectrochemical response. Such modified electrodes form a new field that combines photochemistry, catalytic chemistry, and electrochemistry aimed at developing new photoelectronics.[5,6]

For fabricating modified electrodes, coating by a polymer[5,6] or the Langmuir-Blodgett (LB) membrane[9-14] is a promising approach. There is an important difference between the polymer and LB membranes. Although it may be possible to construct a molecule-based system by utilizing LB membranes, the long alkyl chains of the component could hinder efficient charge transport. In LB systems, integration of the molecule-based process to enhance the output might also be difficult to achieve. Without integration, the photoresponse based on single molecule-based reaction is too small as output for photodevices. In contrast, polymer-coated electrodes have shown that simple integration of the photoreaction components by utilizing multimolecular membranes induces an integrated photoresponse. Although a polymer membrane does not provide ideal molecule-based devices, its utilization must be the first approach to achieving molecular devices when considering its ease of fabrication and integration of reaction components.

Molecule-based photoelectrochemical devices utilizing polymer membranes will be reviewed in this chapter. The typical devices reported are

TABLE 1
Photoelectrochemical Devices Composed of Polymer Membranes

Section	Excitation center	System composed polymer layer Electrode/1st Layer/2nd Layer	Redox reagent in solution	Ref.
III A	$Ru(bpy)_3^{2+}$	BPG/$Ru(bpy)_3^{2+}$/MV^{2+}	None or O_2	50
III A	$Ru(bpy)_3^{2+}$	BPG/$Ru(bpy)_3^{2+}$/PB	—	69
III A	$Ru(bpy)_3^{2+}$	BPG/PB/$Ru(bpy)_3^{2+}$	—	71
III A	$Ru(bpy)_3^{2+}$	ITO/$Ru(bpy)_3^{2+}$/PPhS-BV	None or O_2	61
III B	MPc	ITO/PVCz (ZnPc)/PPhS-BV	None or O_2	78
IV A	$Ru(bpy)_3^{2+}$	ITO/$Ru(bpy)_3^{2+}$	O_2 or MV^{2+} $(+O_2)$	61
IV B	MPc	ITO/PVCz (ZnPc)	O_2 or MV^{2+} $(+O_2)$	78, 80
IV C	MV^+	ITO/MV^+	O_2	83
V B	CdS	CdS/$Ru(bpy)_3^{2+}$ + RuO_2	X^- or H_2O	98, 99
V B	CdS	CdS/Ru-red	H_2O	105
V B	CdS	CdS/$Ru(bpy)_3^{2+}$ + RuO_2 + $Fe(CN)_6^{3-/4-}$	—	106
V C	MPc	ITO/PVCz (ZnPc)	—	78
V A	CdS	Nesa/CdS/PB/comb-type Au/Nafion/PB/Nesa		93
V A	P3MT	ITO/P3MT/PB/aqueous electrolyte/PB/ITO		96

Note: For details, see text.

summarized in Table 1; many of them have been described in earlier reviews,[5,6] so that the major part of this chapter focuses on several recent systems reported by the present authors. The devices described here are divided into the following three types:

1. Photoresponsive devices comprising molecule-based excited state reactions of multilayer polymer membranes.
2. Semiconductor devices functionalized according to molecule-based designs for photocharging and water photolysis purposes.
3. Photoelectrochemical dioxygen sensing based on excited state reactions with dioxygen.

In the next section, basic principles of molecule-based excited state reactions utilizable for photoelectrochemical devices are at first stated briefly followed by the review on these three types of molecule-based devices.

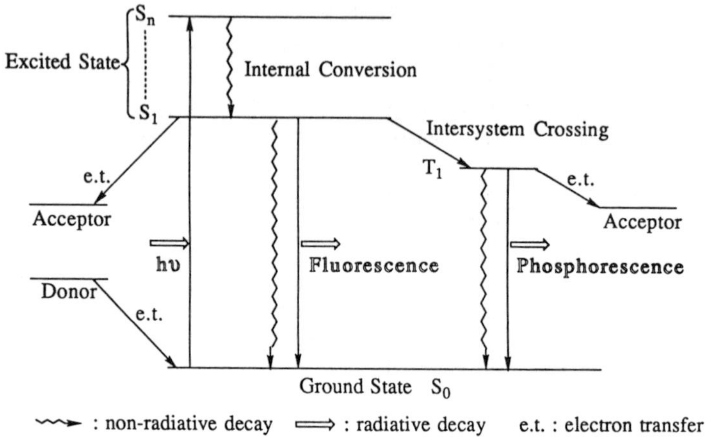

FIGURE 1. Schematic energy diagram showing electron transition from excited state. For details, see text.

II. MOLECULE-BASED ELECTRON TRANSFER REACTIONS

Molecule-based photoelectrochemical devices involve electron transfer reaction at the photoexcited state of a photoexcitation center molecule. When this molecule absorbs photon of suitable energy, the electron is excited from the ground state to the excited state. Since the excited state is a better oxidant and reductant than the ground state, electron transfer can proceed between such an excited state and a suitable reaction partner (electron acceptor or electron donor). This process converts the absorbed light energy into a chemical one.

Basic knowledge on such excited state reactions is important for the design of molecule-based photoelectrochemical devices. For this approach, the following items have to be taken into consideration: (1) general concept about excited state electron transfer; (2) information obtained from excited state electron transfer experiments; (3) modeling of the processes and characterization of devices which utilize the excited state electron transfer reactions.

A. EXCITED STATE ELECTRON TRANSFER REACTION

Various electron transition processes to and from the excited state formed by photon absorption are shown in Figure 1.[15,16] The electron excited by photon from the ground state (S_0) to the singlet excited state ($S_1 \sim S_n$) relaxes to the lowest singlet (S_1) excited state by the internal conversion process. Then, the electron in the excited singlet (S_1) state returns to the thermally equilibrated ground state (S_0) by radiationless transition or by fluorescence emission. If there is a suitable reaction partner (electron acceptor or electron donor) available, an electron transfer reaction may occur between S_1 and the electron acceptor, or between the vacant ground state S_0 and the electron donor. The electron transition from S_1 to excited triplet state (T_1) may occur by intersystem crossing, and the electron in T_1 may encounter the same reactions as those of S_1. The emission from T_1 is called phosphorescence. In addition, isomerization or cleavage of the molecule also can take place from these excited states.

Photoluminescence (fluorescence or phosphorescence) is quenched primarily by an excited state electron or an energy transfer reaction. As we shall see in the next section, luminescence intensities or lifetimes are used to obtain information about the quencher molecules present.

B. PHOTOINDUCED ELECTRON TRANSFER BETWEEN MOLECULES

Photoinduced electron transfer between molecules is receiving more and more attention in conjunction with the study and modeling of highly efficient biological electron transfer reactions, and also with designing artificial photoenergy conversion systems.[17-20] Modeling of processes is especially important for studying the electron transfer in biological molecules. Many of the recent studies used modified biological substances such as cytochrome c and myoglobin as models for studying biological electron transfer between two kinds of redox centers.[21-27] The present authors used amphiphilic zinc porphyrin derivative (ZnP) embedded in liposomes as shown in Figure 2 to study distance effect on the electron transfer from the photoexcited ZnP to methylviologen (MV^{2+}; **1**) present in the outer aqueous phase.[28]

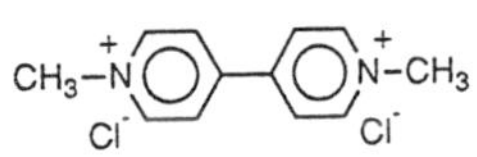

1

The reduction product of the acceptor (MV^{+}) was monitored by a nanosecond laser flash photolysis technique. It was found that intermolecular electron transfer is possible only when the distance between ZnP and MV^{2+} is within 1.2 nm.

Electron transfer from the excited ZnP embedded in liposome to benzoquinone (BQ) added to the outer solution as an acceptor was studied.[29] Analysis of quenching of the excited ZnP gave important information about the incorporation of the acceptor BQ into the liposome as well as the distribution of the incorporated BQ in the liposome. In the MV^{2+} case, it was present only in the outer aqueous phase, and the electron transfer occurred only from the triplet excited ZnP by a diffusion limited value. On the other

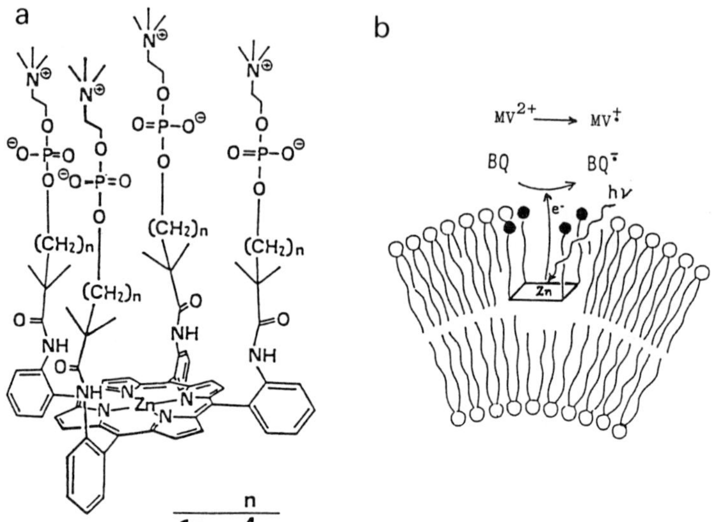

	n
1a	4
b	6
c	10
d	18

FIGURE 2. (a) Amphiphilic zinc porphyrin (ZnP) and (b) liposome of the ZnP with phospholipid.

hand, BQ reacted with both the singlet and triplet excited states of ZnP, indicating that the BQ is incorporated into the liposome and causes static reaction of the singlet excited ZnP. From the equilibrium constant of the incorporation correlated with the relative emission yield, the distribution of the incorporated BQ in the liposome could be obtained.[29]

C. EXCITED STATE REACTIONS UTILIZED FOR PHOTOELECTROCHEMICAL DEVICES

Almost all the photodevices available at present are based on photophysical processes of semiconductors, and are called photovoltaic systems.[30] Photodevices could also be constructed by utilizing excited state electron transfer reactions (photochemical processes). These photodevices could be called photogalvanic systems.[6]

Electron transfer at the photoexcited state can be classified into two types. One is a so-called photoredox system in which the equilibrium under illuminated conditions is shifted to the photochemical product side because the rate of the back electron transfer (v_b) is slower than the forward electron transfer (v_e) at the excited state (Equation 1):

$$P + A \xrightarrow{h\upsilon} P^* + A \underset{v_b}{\overset{v_e}{\rightleftharpoons}} P^+ + A^- \quad (1)$$

$$\text{Photoredox type:} \quad v_e > v_b$$
$$\text{Nonphotoredox type:} \quad v_e < v_b$$

A typical example of the photoredox system is the coupling of thionine dye and ferrous ion in aqueous solution.[31-45] Another type is a nonphotoredox system in which the equilibrium under illuminated conditions is not shifted to the product side because rate of the back electron transfer is comparable to, or even more rapid than, the forward electron transfer. Most of the excited state electron transfer reactions are the nonphotoredox type. Photogalvanic cells can be fabricated simply by dipping electrodes into the aqueous solution of the photoredox reaction couple. However, photoelectrochemical cells could not be simply fabricated by utilizing nonphotoredox systems present in solution. The present authors have found that when the electrode is coated with one or both the components of the photoreaction couple, photoresponse can be obtained at the coated electrode even by utilizing the nonphotoredox reaction couple. This is because the dynamics of the electron transfer process between the electrode and the photochemical reaction products may overcome the rapid back electron transfer reaction of the products. The confinement of at least one of the reaction components onto the electrode is the key point for constructing photoelectrochemical devices based on nonphotoredox-type electron transfer at the excited state. The possibility of developing photodevices based on the photoredox couple seems to be limited because of the limited number of known photoredox couples. Such systems have been reviewed in Kaneko and Wöhrle[6] in detail so that in this chapter only the photodevices based on a nonphotoredox couple will be described.

A typical example of a nonphotoredox reaction is electron transfer from the excited state (T_1) of *tris*(2,2′-bipyridine)ruthenium(II) complex (abbreviated to $Ru(bpy)_3^{2+}$; **2**) to MV^{2+}:

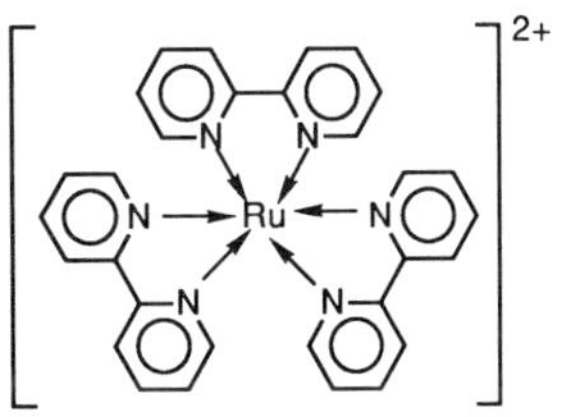

2

This couple has been studied intensively as a candidate for a photoreaction center aiming at water photolysis.[5] The rate constant of the electron transfer from the excited Ru complex to MV^{2+} in an aqueous solution is the order of

10^8 $dm^3/mol/s$ (which is near the diffusion-limited value) while that of the back electron transfer, e.g., from MV^+ to $Ru(bpy)_3^{3+}$, is about twice that of the forward electron transfer. Any photoresponse cannot be obtained at an electrode dipped into the mixed solution of these photoreaction components. When the Ru complex is confined on the electrode, however, photoresponse is obtained at the modified electrode with MV^{2+} present either in the aqueous phase[46-49] or as the second polymer layer coated on top of the Ru complex layer.[50]

Metal complexes of macrocyclic ligands such as metalphthalocyanines (MPc; **3**) and metal porphyrins (MP; **4**) are also good candidates for the photoexcitation center:

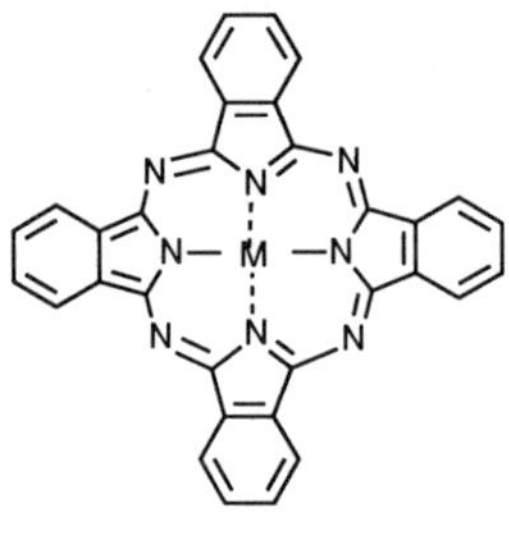

3

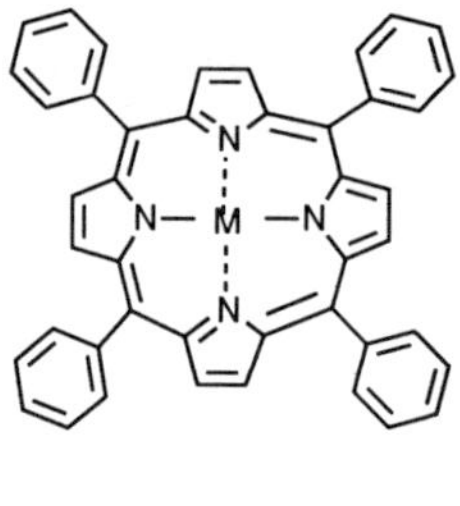

4

Refer also to Table 1 in which the combination of photoexcitation center and acceptor molecules reported by the present authors are briefly summarized. As we shall see in the following sections, organic and inorganic semiconductors can be combined with various functional molecules by utilizing polymer coating to realize new functionalized photodevices based on molecular reactions such as photochargeable devices and water photolysis devices.

III. PHOTOELECTROCHEMICAL DEVICES COMPRISING MOLECULE-BASED EXCITED STATE REACTIONS OF MULTILAYER MEMBRANES

As described before, almost all the photoresponsive devices are based on the photophysical process of semiconductors, but new photodevices could be constructed based on molecular electron transfer of a photoexcited state of a light-absorbing compound by immobilizing at least one of the reaction components on the electrode. This is a crucial point for obtaining photodevices based on a nonphotoredox couple, since in a homogeneous solution the unavoidable rapid back reaction brings about no net response to the exterior. Fabrication of a multilayer structure of thin polymer membranes containing

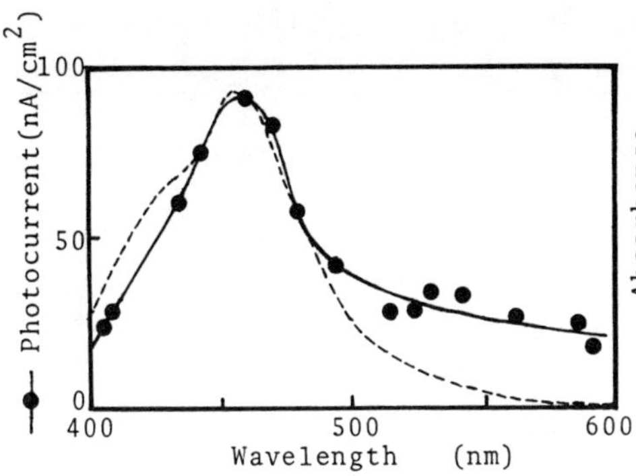

FIGURE 3. Action spectra for the photocurrent obtained at BPG electrode coated with Ru[poly(St-Vbpy)](bpy)$^{2+}$. Measured in 20 m*M* MV^{2+} aqueous solution (0.2 *M* CF_3COOH, pH 2) at the electrode potential of −0.2 V (vs. Ag-AgCl) under argon.

each reaction component is a better solution to this problem. In this section, photoresponsive devices utilizing excited state reactions of multilayer polymer membranes are described.

A. Ru(bpy)$_3^{2+}$ AS A PHOTOEXCITATION CENTER

The present authors have reported the generation of photocurrent at a graphite electrode coated with polymer-pendant Ru(bpy)$_3^{2+}$ (**5**) membrane in the presence of MV^{2+} in the aqueous solution in contact with the coated electrode.[46,48]

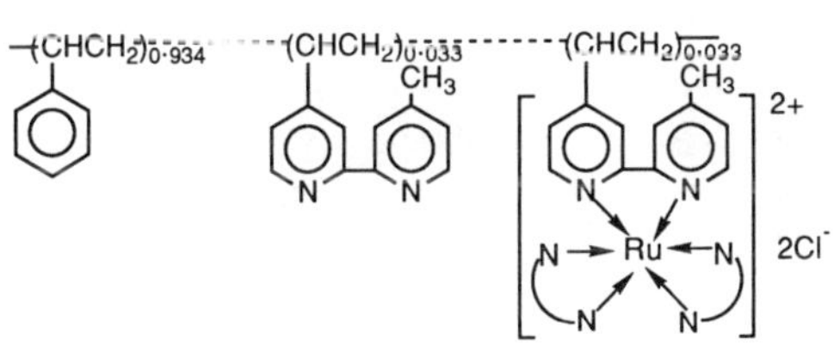

5

The action spectrum of the photocurrent (Figure 3) showed that excitation of the Ru complex is responsible for the photoelectrochemical event.

In this system composed of an electrode/Ru(bpy)$_3^{2+}$/MV^{2+} arrangement, the photocurrent should be cathodic if the arrangement and the direction of the electron flow are ideal. However, because of the penetration of MV$^+$ into the polymer layer containing Ru(bpy)$_3^{2+}$, anodic photocurrent was also observed when the electrode was polarized positively. An anisotropic photoinduced electron flow is important in producing efficient photocurrent. Anisotropic arrangement of the reaction components is necessary for this purpose. Utilization of multilayer polymer membranes is one of the promising ways to achieve such anisotropic arrangement.

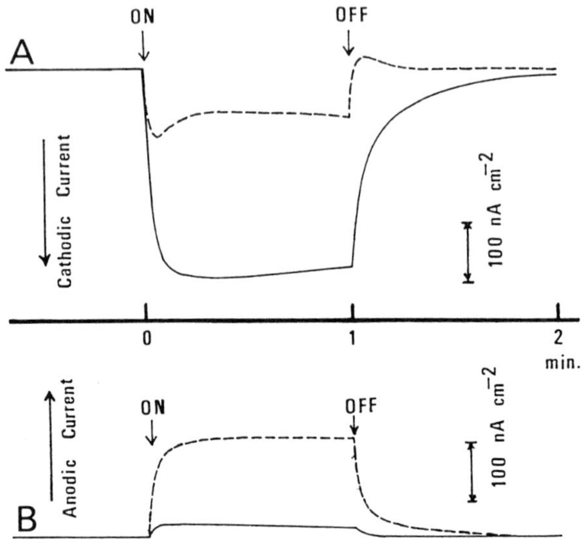

FIGURE 4. Photocurrents induced at a {BPG/Ru[poly(St-Vbpy)](bpy)$^{2+}$/poly(4-MV)$^{2+}$} bilayer coated electrode (solid line) and at a [BPG/mixture of two polymers] monolayer coated electrode (dashed line) dipped in 0.2 *M* CF_3COONa aqueous solution adjusted to pH = 2.0. Photocurrents were measured at −0.2 V (A) and +0.7 V (B).

The arrangement of $Ru(bpy)_3^{2+}$ and MV^{2+} was attained by coating a graphite electrode (basal plane pyrolytic graphite [BPG]) at first with a polymer pendant Ru complex membrane and then with a polymer pendant MV^{2+} (**6**) membrane.

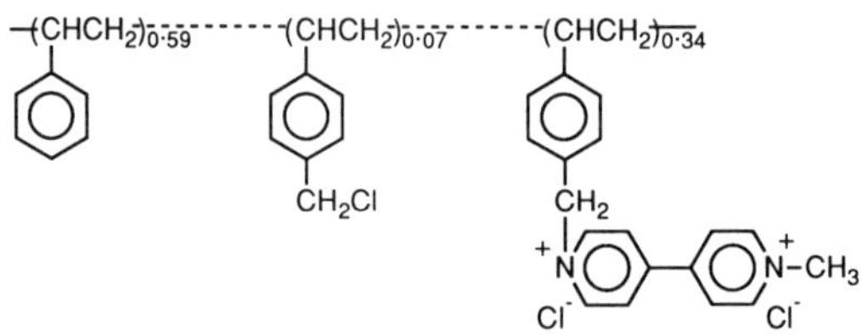

6

The time-current behaviors of the bilayer-coated system induced by irradiation and cutoff of the light under argon are shown in Figure 4 in comparison with those obtained for a mixture-coated electrode.[47,50] The cathodic photocurrent generated by the bilayer-coated system at the electrode potential of −0.2 V (vs. Ag-AgCl) is four times higher than that by the mixture-coated one. The anodic photocurrent obtained at 0.7 V is, on the contrary, 13 times higher for the mixture coating than for the bilayer coating. The effect of the bilayer system to produce anisotropic photocurrent is calculated to be 52 times

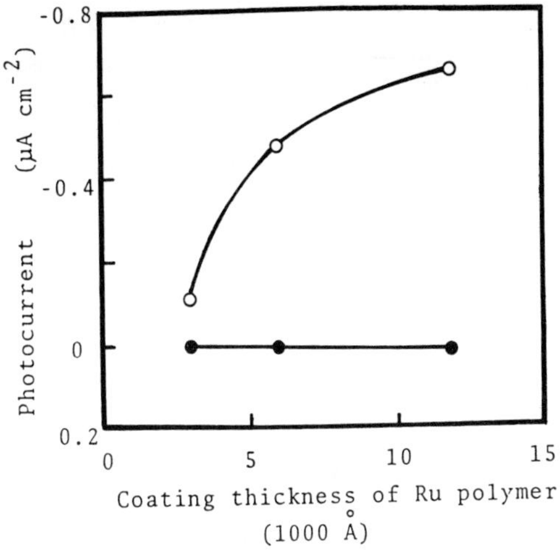

FIGURE 5. Dependences of photocurrent on coating thickness of a bilayer (**5** and **6**) coated electrode dipped in 0.2 M CF_3COOH adjusted to pH 2. Applied potential of −0.19 V (○) and 0.61 V (●) vs. Ag-AgCl, measured under argon. The molar ratio of Ru/MV^{2+} is taken constant (0.55).

as high as the mixture-coated system. The small anodic photocurrent at 0.7 V for the bilayer system indicates, however, that the electrode reaction of MV^+ also occurs to a small extent. It would be due to a nonideal bilayer structure, e.g., penetration of the viologen polymer into the $Ru(bpy)_3^{2+}$-coated layer. The dependence of the cathodic and anodic photocurrents on the membrane thickness for this bilayer-coated system is shown in Figure 5. Selective cathodic photocurrent was induced that increases with increasing membrane thickness, while the small anodic photocurrent is almost independent of the thickness. These results show that the thicker coating hinders the electrode reaction of the viologen moieties more, leading to higher cathodic photocurrent and higher light-to-electricity conversion efficiency.

For the bilayer-coated system, electrode/polymer-pendant $Ru(bpy)_3^{2+}$/polymer-pendant MV^{2+}, the electron flux of each step has been estimated as shown in Figure 6.[6] The electron transport in the polymer membrane occurs by electron exchange or hopping between Ru complexes or between MV^{2+} moieties. The fluxes due to the photoreaction are very large, while the electron transport inside of the membrane is the slowest process among the photoelectrochemical processes concerned. This process occurring by electron exchange between the Ru complexes is regarded as a kind of diffusion of the charge. Under this circumstance, the overall conversion efficiency is not high (the order of 0.05% based on the total incident monochromatic light) because the energy-consuming back electron transfer prevails still in the present system. When considering the scheme of Figure 6, the conversion efficiency

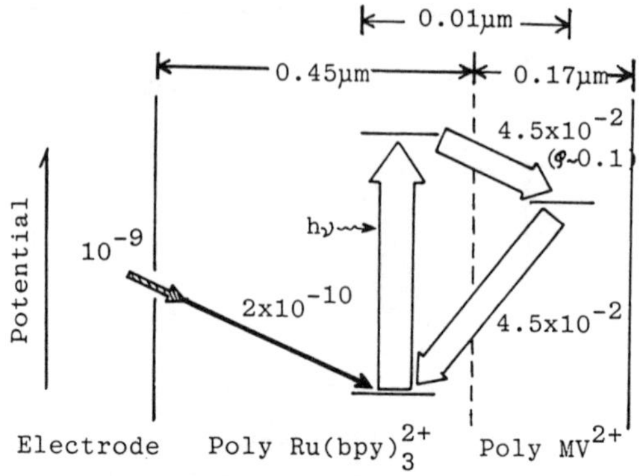

FIGURE 6. Estimation of electron flux in photoelectrochemical processes at the bilayer (**5** and **6**) coated electrode. Units of the values: $mol/cm^2/s$.

should be higher for the thinner membrane systems since the distance of the rate-determining slow electron transport process is shorter. This should be true only when assuming an ideal arrangement of the components. However, a thicker membrane actually gave a higher photocurrent as shown in Figure 4. This result would indicate that the two components, Ru complex and viologen, are not arranged ideally, but are mixed with each other at the interface so that the penetrated MV^{2+} still has a chance to react at the electrode in its reduced form resulting in the decrease of the cathodic photocurrent. Thicker Ru complex membrane produced higher cathodic photocurrent probably by preventing this unfavorable reaction and also by integration effect of the light-absorbing Ru complex.

It is interesting that electron transfer can occur in the polymer membrane that is essentially insulating rather than conductive with respect to charge transport. It has been established that charge can propagate by exchange (or hopping) between redox centers in this kind of coated membrane.[51-54] In this charge propagation system, the distance between the redox centers should be taken into account since a maximum distance should exist for the charge to be transported as described in Section II.B. The distance between redox centers dispersed in a membrane was calculated by a statistical treatment.[55,56] The probability of a redox center present in a small sphere obeys the Poisson distribution.[29,57,58] The probability density [P(r)] of the intermolecular distance (r) between the redox centers is calculated from this Poisson treatment as follows:

$$P(r) = 4\pi r^2 N_A c \exp(-4\pi r^3 N_A c/3) \tag{2}$$

where N_A is Avogadro's number, and c the complex concentration in the membrane bulk. The factor of $\exp(-4\pi r^3 N_A/3)$ represents the probability of

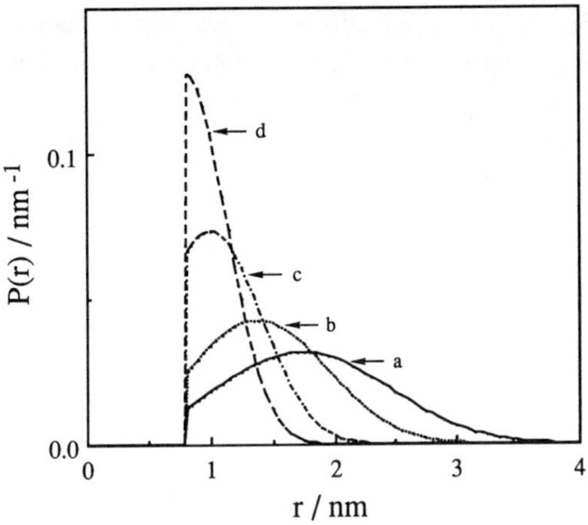

FIGURE 7. Probability distribution of intermolecular distance (r) of the $Ru(bpy)_3^{2+}$ redox center dispersed in a polymer membrane. $Ru(bpy)_3^{2+}$ concentration in mol/dm^3: (a) 0.05, (b) 0.1, (c) 0.261, and (d) 0.5.

a central redox center in the sphere around which (radius r) no other redox center is present, and the factor $4\pi r^2 N_A c$ the probability of a redox center in the sphere around which another redox center is present. In addition to Equation 2, the excluded volume of the redox center (radius a = 0.8196 nm for $Ru(bpy)_3^{2+}$ as estimated from crystallographic analysis of the Ru complex) should be considered as Equation 3:

$$P(r) = 4\pi r^2 N_A c \exp\{-4\pi(r^3 - a^3)N_A c/3\} \quad (3)$$

According to Equation 3, the probability distribution of intermolecular distance for the redox center of $Ru(bpy)_3^{2+}$ dispersed homogeneously in a polymer membrane is calculated as shown in Figure 7 under various concentrations of the Ru complex. In the present polymer complex (**5**), the concentration of the Ru complex of the film is calculated to be 0.261 mol/dm^3 when assuming that the density of the coated polymer is 1 g/ml. The distance between the Ru redox centers which gives the maximum probability in Figure 7 is obtained as 1.0 nm for polymer 5. This result shows interestingly that electron transfer in this insulating polymer membrane can occur by electron hopping between the Ru complex, when considering the finding for the intermolecular electron transfer described in Section II.B that electron can be transferred within 1.2 nm.[28]

As for the polymer backbone to incorporate redox centers, we were interested in polysiloxanes. Polysiloxanes have a flexible main chain of –Si–O– bond, which allows interaction of pendant functional groups themselves or interaction between pendant functional groups and substrates present in the liquid phase. The affinity of polysiloxane to inorganic materials such as glass and carbon is very high.[59] From these points of view, a new polysiloxane-pendant viologen was prepared by the reaction of chloromethylated poly(methyl-*p*-phenethylsiloxane) with monoalkylbipyridinium chloride in dry dimethyl formamide (DMF) to obtain polysiloxane-pendant benzylviologen (PPhS-BV; **7**).

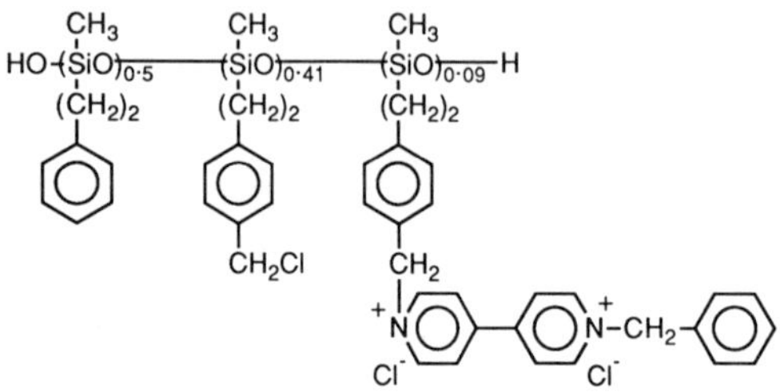

7

The film of PPhS-BV coated on indium tin oxide (ITO) was stable, and the reduced viologen groups are very sensitive to oxidation by oxygen as revealed by voltammetric studies.[60] This vologen polymer was applied to photoresponsive devices.

An ITO electrode was at first coated with a thin film of an anionic polymer such as Nafion (thickness: 2.5 μm), and then cationic $Ru(bpy)_3^{2+}$ was adsorbed into the polyanion layer electrostatically from aqueous solution. A bilayer-coated electrode (ITO/Nafion[$Ru(bpy)_3^{2+}$]/PPhS-BV) was prepared by casting viologen-pendant polysiloxane solution in methanol on top of the ITO/Nafion[$Ru(bpy)_3^{2+}$] membrane.[61] This bilayer-coated device produced only cathodic photocurrent independent of the applied potential indicating that the arrangement of the components is more ideal than the system described above. In this system, photocurrent was especially enhanced by oxygen present as a second acceptor because of the high sensitivity of the reduced polysiloxane-pendant viologen toward oxygen. This is described later again in Section IV.

In a water photolysis system composed of an aqueous solution of colloidal Prussian blue (PB) and $Ru(bpy)_3^{2+}$, photoinduced electron transfer was concluded to occur from the excited Ru complex adsorbed on the PB colloidal particle to PB.[62,63] PB is a mixed valent iron cyanide complex with the repeating composition:

$$K^+Fe^{3+}[Fe^{II}(CN)_6]^{4-} \qquad (\mathbf{8}\text{, soluble PB; see Figure 8})$$

which forms a high molecular polynuclear structure.[57,64,65] The photochemical reaction of PB and the Ru complex must also be applicable to fabricating photoresponsive devices. A bilayer system coated with $Ru(bpy)_3^{2+}$ and PB for an electrochemical device in addition could give information on the photo-

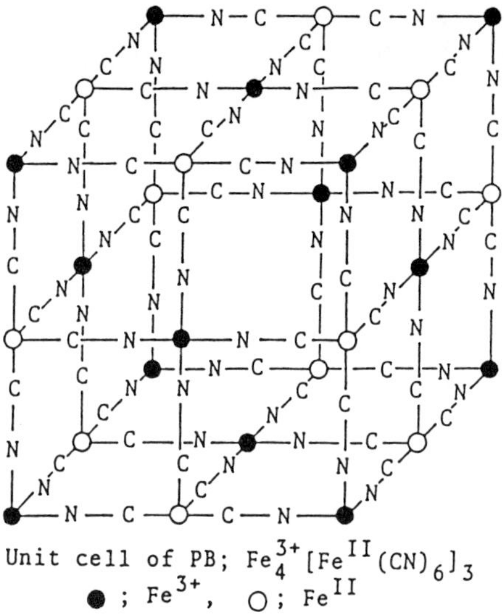

FIGURE 8. Illustrative depiction of the unit cell of insoluble form of PB (**9**). The soluble PB (**8**) has a structure with additional $Fe^{II}(CN)_6$ repeating unit in the center of this cell.

chemical reaction between the components. A graphite BPG electrode was at first coated with a polymer pendant $Ru(bpy)_3^{2+}$ membrane from its DMF solution, and then with a PB membrane by reductive electrodeposition from the aqueous colloidal Berlin brown (BB) solution–Fe^{3+}-Fe^{3+} complex, formed by mixing $[Fe(CN)_6]^{3-}$ and Fe^{3+} in water.[66,67] The electrodeposited PB called an insoluble PB has the repeating unit:

$$Fe_4^{3+}[Fe^{II}(CN)_6]_3^{4-} \qquad (\mathbf{9}\text{, insoluble PB; see Figure 8})$$

whose unit cell is shown in Figure 8.[68] This complex actually forms a very high molecular weight (7 million) compound.

The bilayer-coated system:

$$BPG/polymer\ Ru(bpy)_3^{2+}/PB$$

gave mainly cathodic photocurrent.[69] The action spectra for the photocurrent showed that for the cathodic photocurrent the major process is electron transfer from the electrode to PB via the excited state of the Ru complex. Such a photoelectrochemical process supported the mechanism of water photolysis.[62,63] Direct electron transfer from the electrode to the excited PB is the minor process for the cathodic photocurrent for which the Ru complex is working only as a charge transport layer. This direct photoelectrochemical process between the electrode and the excited PB was confirmed by a separate experiment using a PB-coated electrode.[70]

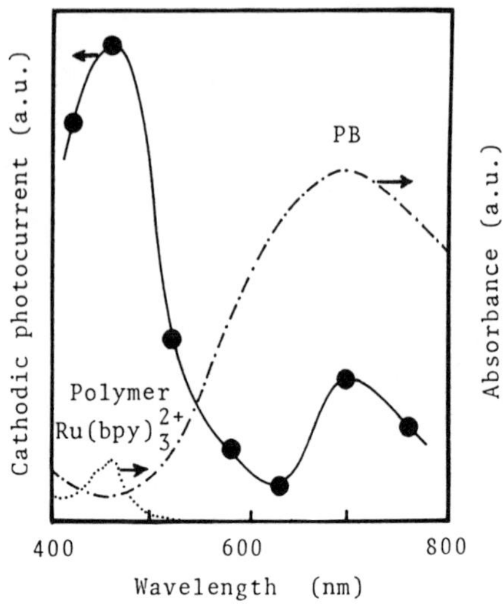

FIGURE 9. Action spectra of cathodic photocurrent obtained for {BPG/PB/Ru[poly(St-Vbpy)](bpy)$^{2+}$} bilayer (**9** and **5**) coated electrode at the applied potentials of −0.3 V vs. Ag-AgCl.

The bilayer-coated system composed of an arrangement of the components reverse to the above system:

$$\text{BPG/PB/polymer Ru(bpy)}_3^{2+}$$

also produced a predominant cathodic photocurrent.[71] The action spectrum for the cathodic photocurrent obtained under cathodic polarization (Figure 9) showed that the major process is the electron transfer from the electrode to the vacant ground state of the Ru complex — Ru(bpy)$_3^{2+}$ — via the reduction of PB (Figure 10a). The minor photoelectrochemical process for the cathodic photocurrent is the direct electron injection from the electrode to the excited PB (Figure 10b).

A considerable number of electron donors or acceptors can be used as reaction partners for the excited Ru(bpy)$_3^{2+}$ to fabricate photoelectrochemical devices, which is a problem to be studied in the future.

B. METAL PHTHALOCYANINES AS A PHOTOEXCITATION CENTER

Thin membranes of various phthalocyanines (Pc's) have been investigated as sensitizers for photovoltaic devices composed of *n*-type inorganic semiconductors in contact with an aqueous solution containing a redox couple, such as I_2/I^-, Fe(CN)$_6^{3-/4-}$, *p*-benzoquinone/hydroquinone, or Fe(III)/Fe(II)-ethylenediaminetetraacetic acid.[6,72-76] Phthalocyanines also have often been

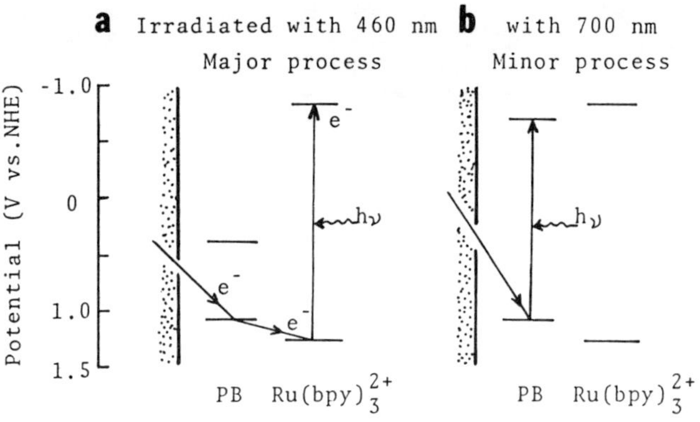

FIGURE 10. Mechanism of cathodic photocurrent generation at bilayer (**9** and **5**) coated BPG. (a) Major process, (b) minor process.

used as *p*-type semiconducting materials for fabricating photovoltaic devices. Dispersion of Pc in a polymer binder increases the conversion efficiency by bringing about a better separation of electron-hole pairs formed under irradiation.[76,77]

Metal phthalocyanines (MPc's) also can be used as a photoexcitation center for fabricating molecule-based photodevices.[6] Recently it has been found that, for the MPc photoexcitation center, oxygen is especially effective as a primary electron acceptor from the excited MPc. Such a system is described in Section IV. In this section, bilayer-coated systems composed of ZnPc dispersed in poly(*N*-vinylcarbazole) (PVCz) or in polysiloxane (PPhS) and polysiloxane-pendant viologen will be briefly described. The devices, ITO/PVCz[ZnPc]/PPhS-BV and ITO/PPhS[ZnPc]/PPhS-BV, generated mainly cathodic photocurrent.[78] The oxygen present enhanced the photocurrent to a great extent, which will be described in Section IV. MPc alone did not induce any photocurrent without an acceptor.

IV. PHOTOELECTROCHEMICAL DEVICES BASED ON EXCITED STATE REACTIONS WITH DIOXYGEN

Dioxygen is often an efficient electron acceptor for a photoexcited state of a compound. This fact suggests utilization of dioxygen as a primary acceptor for an excited state of photoreaction center to fabricate photoelectrochemical devices based on molecular electron transfer. Dioxygen is also an efficient acceptor for reduced acceptor molecules such as viologen and quinones, which suggests an enhanced photoresponse induced by photoelectrochemical devices in the presence of dioxygen. In addition to this, such photoelectrochemical devices operated in the presence of dioxygen would lead to a new dioxygen sensing system as a form of photoelectrochemical sensor. From these view-

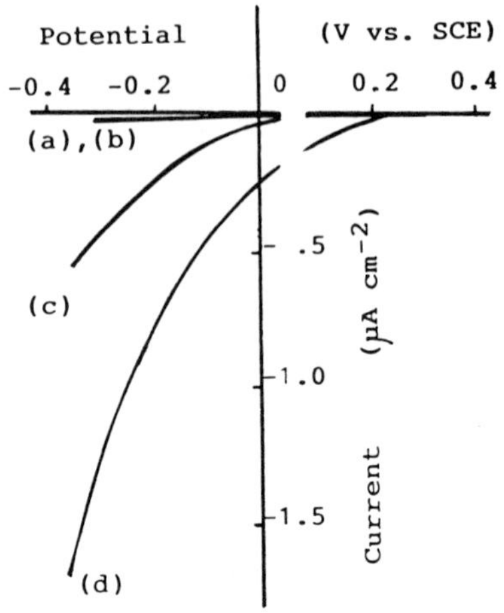

FIGURE 11. Dependence of current changes on the applied potential at ITO/Nafion[Ru(bpy)$_3^{2+}$] dipped in a 0.1 *M* KCl aqueous solution. (a) In the dark, under argon; (b) under irradiation, under argon; (c) in the dark, under air; (d) under irradiation, under air.

points, it is an important and interesting subject to study the effect of oxygen as an acceptor in photoelectrochemical devices based on molecular electron transfer. In this section, photoelectrochemical devices composed of a photoexcitation center and operated in the presence of O_2 will be described.

A. Ru(bpy)$_3^{2+}$ as a Photoexcitation Center

It is known that O_2 accepts electrons from the photoexcited Ru(bpy)$_3^{2+}$. This photoinduced electron transfer causes quenching of the phosphorescence from the photoexcited Ru complex, and has been successfully applied for the quantitative analysis of O_2 by utilizing a polymer membrane incorporating an Ru(bpy)$_3^{2+}$ photoluminescent probe.[15,79] Such a photoinduced excited state reaction must be applicable also to photoelectrochemical devices based on an excited state reaction with O_2. It has been demonstrated for the first time that an ITO electrode coated with a polymer membrane (Nafion) containing Ru(bpy)$_3^{2+}$ induces photocurrent when dipped in an aqueous electrolyte solution in the presence of O_2 (as shown in Figure 11).[61] The action spectrum showed that photoexcitation of the complex induced photocurrent in the presence of O_2, which is good evidence that O_2 works as an acceptor for the excited Ru complex to induce photocurrent. Coating of polymer-pendant viologen on the top of the complex layer enhanced much the photocurrent, for which the viologen works as an electron mediator from the excited complex to O_2.[61,78]

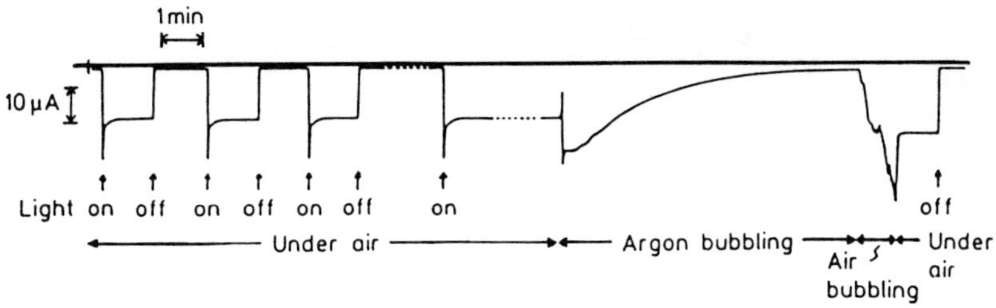

FIGURE 12. Current changes induced by switching on and off of the irradiation, and by argon or air bubbling.

B. METAL PHTHALOCYANINES AS A PHOTOEXCITATION CENTER

As described in Section III, photocurrent was observed in a monolayer ITO/PVCz[ZnPc] system as well as in a bilayer ITO/PVCz[ZnPc]/PPhS-BV system when dipped into an air-saturated aqueous solution. While the monolayer ITO/PVCz[ZnPc] device could not induce photocurrent in the absence of O_2, the bilayer device — ITO/PVCz[ZnPc]/PPhS-BV — generated photocurrent also in the absence of O_2 although the photocurrent was enhanced remarkably in the presence of O_2.[78]

Figure 12 shows current changes at ITO/PVCz[ZnPc] dipped in an aqueous electrolyte solution induced by switching the irradiation on and off and by argon or air bubbling into the cell solution.[80] This result shows that oxygen works as an acceptor to induce cathodic photocurrent. The cathodic photocurrent was dependent on the oxygen concentration and tended to saturation after about 0.5 m*M* concentration (Figure 13a).[78] This saturation tendency could be analyzed when assuming complex formation of O_2 with ZnPc followed by electron transfer from the excited ZnPc to O_2 (Equation 4):

$$\mathrm{ZnPc} + \mathrm{O_2} \underset{k_2}{\overset{k_1}{\rightleftarrows}} \mathrm{ZnPc} - \mathrm{O_2} \xrightarrow[h\upsilon]{k_3} \mathrm{ZnPc^+} + \mathrm{O_2^-} \tag{4}$$

Similar to the Michaelis-Menten type of mechanism of enzymatic reactions, the Lineweaver-Burk-type Equation 5 is derived from Equation 4, where v is the electron flux in mol/cm/s, v_m is its maximum value represented by $v_m = k_3[\mathrm{ZnPc}]_o$, and K_m is the so-called Michaelis constant represented by $K_m = (k_2 + k_3)/k_1$:

$$1/v = (K_m/V_m)(1/[O_2]) + 1/V_m \tag{5}$$

The plots of 1/v vs. $1/[O_2]$ give a linear relationship (Figure 13b) supporting the complexation mechanism of O_2 with the metal complex. From Figure 13b,

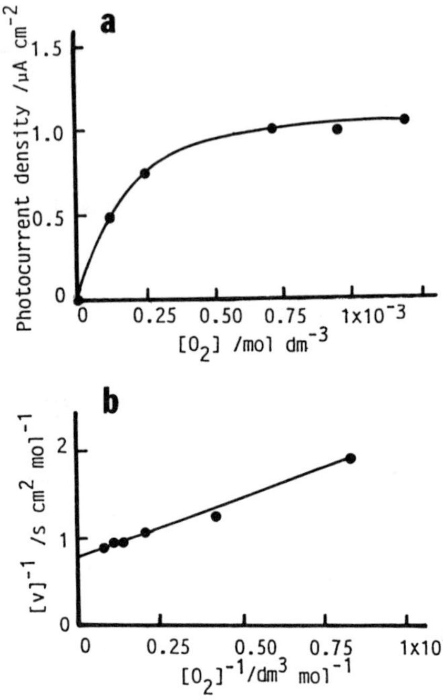

FIGURE 13. (a) Dependence of photocurrent on $[O_2]$ at ITO/PVCz[ZnPc] coated electrode dipped in a 0.1 *M* KCl aqueous solution. Applied potential, 0 V vs. SCE. (b) Lineweaver-Burk plots based on Equation 5.

the constant for the complex formation ($K_1 = 1/K_m$) was calculated as 6.06 $\times 10^3$ (M^{-1}). Thus, the O_2 molecule is considered to work as an acceptor by coordinating to the metal complex rather than by physical diffusion into the polymer film.[78]

In this type of photoelectrochemical device, a pure MPc layer prepared by vapor evaporation did not give good results. It was important to disperse the complex into a polymer membrane to coat an electrode. Various polymers were used for this purpose such as polystyrene (PS), poly(acrylonitrile) (PAN), poly(1-vinylcarbazole) (PVCz), or poly(vinylidene fluoride) (PVF), etc.[81,82] The nature of the polymer had a strong influence on the photocurrent. The polymer with higher dielectric constant gave higher photocurrent as shown in Figure 14. It is suggested that higher dielectric constant would provide better environment around the MPc for better electron/hole separation as well as charge transport.

MPc has a *p*-type semiconductor character, and some data suggested that the mechanism might be better explained based on semiconductor rather than based on molecular photochemistry. Detailed studies are still under way.[82]

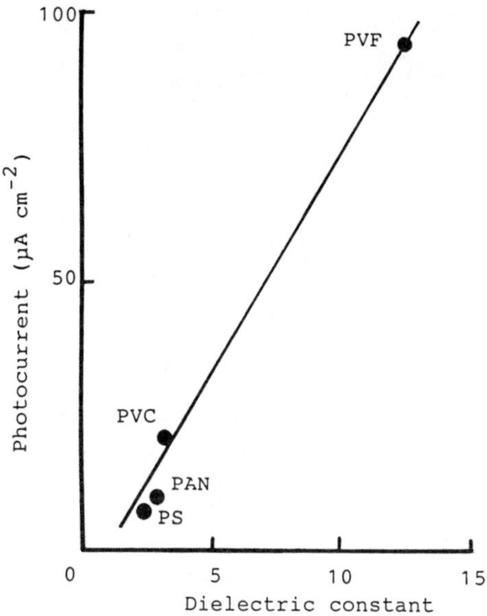

FIGURE 14. Effect of dielectric constant of polymer on the photocurrent generated by ZnPc/polymer membrane dipped in aqueous electrolyte.

C. ELECTROCHEMICAL PRODUCT WORKING AS AN ELECTRON-DONATING PHOTOEXCITATION CENTER FOR AN O_2 ACCEPTOR

MV^{2+} has been studied intensively as an acceptor or electron carrier in a number of photochemical and electrochemical reactions.[1-8] An entirely new photogalvanic effect of MV^{2+} cation radical (MV^+) formed electrochemically on the surface of an ITO electrode was found recently with an O_2 acceptor. A cation radical of MV^{2+} formed on the surface of an ITO electrode dipped in an aqueous electrolyte solution gave cathodic photocurrent in the presence of oxygen.[83]

The formation of MV^+ under negative applied potentials was confirmed by a visible color formation as well as by a visible absorption spectrum. In the absence of oxygen, it was observed that the MV^+ formed at the ITO surface diffused slowly into the bulk aqueous phase. In the presence of O_2, MV^+ was observed only on the ITO surface since the MV^+ diffusing into the solution is subjected to rapid oxidation by O_2. Under the applied potentials (-0.8 V vs. SCE) in the presence of O_2, the irradiation of MV^+ formed on the ITO surface generated cathodic photocurrent. The photocurrent increased almost linearly with the light intensity between 50 ~ 120 mW/cm^2, but no photocurrent was induced below 50 mW/cm^2. The mechanism of the photocurrent generation was ascribed to the reaction of the excited state MV^+

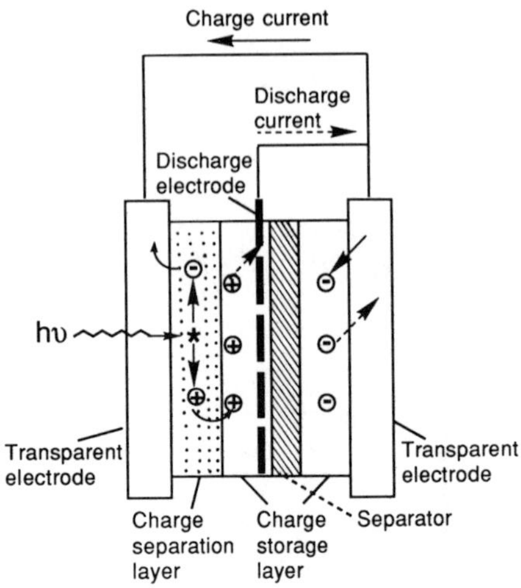

FIGURE 15. Conceptual illustration of photochargeable device composed of multilayer membranes.

with oxygen. The excited state MV^+ would have a much higher reactivity toward oxygen than the ground state MV^+, which would cause cathodic photocurrent.

V. PHOTOFUNCTIONAL DEVICES BASED ON MOLECULE-BASED REACTIONS

Various photofunctional devices can be designed based on molecule-based reactions. For instance, a photochargeable device capable of charging under irradiation and discharging in the dark based on molecular reaction could open a new field of photoenergy conversion and storage. Such a device also can be used for information storage. Some model devices for photoenergy conversion and storage have been constructed by using liquid-junction inorganic and organic semiconductors in contact with a redox electrolyte solution.[84-88] Photodevices for photoenergy conversion are also important subjects for research relevant to new energy resources. To realize such devices it is also significant to adopt molecule-based reactions. Multilayer polymer membranes with each layer possessing various functions provide important materials for this purpose. In this section, photochargeable devices and photoenergy conversion devices will be described which utilize polymer membranes functionalized with incorporated molecules.

A. PHOTOCHARGEABLE DEVICES

The concept of a photochargeable device composed of multilayer membranes is shown in Figure 15. Each layer bears a function such as photoinduced

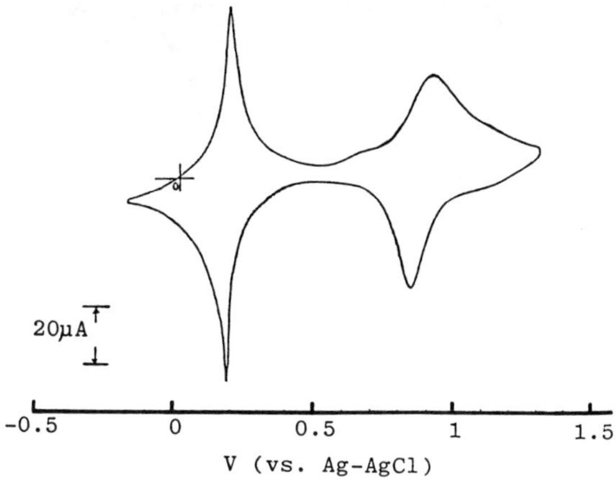

FIGURE 16. Cyclic voltammogram of ITO/PB coated electrode measured in 0.1 *M* KNO_3 and 0.01 *M* HNO_3 aqueous electrolyte.

charge separation, accumulation of positive/negative charges, and separation of reactants. As for the electrode, a transparent ITO is suited. For photoinduced charge separation, a semiconductor is the first promising candidate which forms a kind of Schottky junction (liquid junction) when in contact with an electrolyte solution. To accumulate charges, the present authors are interested in Prussian blue (see Section III.A), since it is capable of storing both positive and negative charges. Since PB is an interesting material for charge storage,[89-91] its characteristics are described next.

As stated before, PB is a mixed valent iron cyanide complex with the composition of **9** (see also Figure 8). This blue-colored complex can be deposited electrochemically on an electrode as a thin membrane (as shown before). The electrochemical behavior of the PB membrane has been well characterized.[64-66] Since PB can be both reduced and oxidized reversibly (refer to cyclic voltammogram shown in Figure 16), it can store both negative and positive charges, allowing its use as an active material for both the anode and the cathode of a secondary battery. Thus, a semidry multilayer-type battery was constructed by sandwiching a Nafion film between a pair of PB-coated electrodes. The multilayer PB secondary battery:

ITO/PB/Nafion/PB/ITO

showed stable charge/discharge characteristics.[92] The redox reactions of PB membranes accompanying the charge/discharge processes were studied by a spectroelectrochemical technique. On charging, the anodic PB is oxidized to form Berlin brown (BB; Fe^{3+}-Fe^{3+} complex) and the cathodic PB reduced to form Prussian white (PW; Fe^{2+}-Fe^{2+} complex). On discharging, both the

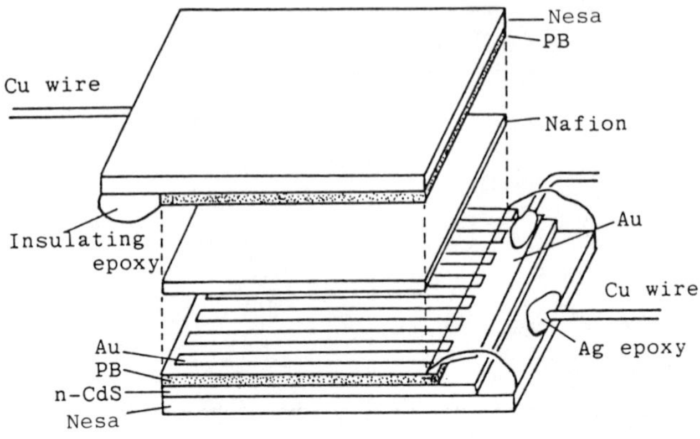

FIGURE 17. Photochargeable device comprising multilayer membranes: Nesa/n-CdS/PB/Au/Nafion/PB/Nesa.

membranes recover their original PB states, thus making repeated charge/discharge cycles possible. This PB secondary battery was utilized as a charge storing material in photochargeable devices.

A multilayer photochargeable device driven by visible light irradiation has been fabricated which is composed of a polycrystalline n-CdS film and PB battery membranes. The photoelectrochemical properties of the polycrystalline n-CdS film coated on Nesa (SnO_2) glass by a chemical spray deposition method was characterized well before utilizing for the photochargeable device.[93] PB was coated on the top of the CdS film by electrodeposition for positive charge storage. The counter electrode was a PB-coated Nesa glass used for negative charge storage. The photochargeable multilayer device was fabricated by sandwiching a Nafion film (173-μm thickness) adsorbing aqueous electrolyte (0.5 M KNO_3 and 10 mM HNO_3) with a PB-coated CdS film/Nesa and a PB-coated Nesa. For obtaining a device capable of discharge in the dark after photocharge, a comb-type Au electrode was attached on the PB film by vapor deposition before sandwiching. The device composition shown in Figure 17 is

Nesa/CdS/PB/comb-type Au/Nafion/PB/Nesa

The area of the effective cell surface was about 0.20 cm^2. This device showed photocharging and dark discharging abilities as illustrated in Figure 18. Such photocharge and dark discharge cycles could be repeated, although the dark discharge/photocharge ratio (73%) was worse than the discharge/charge ratio (90%) of the PB/Nafion/PB battery mentioned earlier.

An organic semiconductor such as poly(thiophene) (PT) could also be used for photoinduced charge separation. Photocurrent generation by a liquid

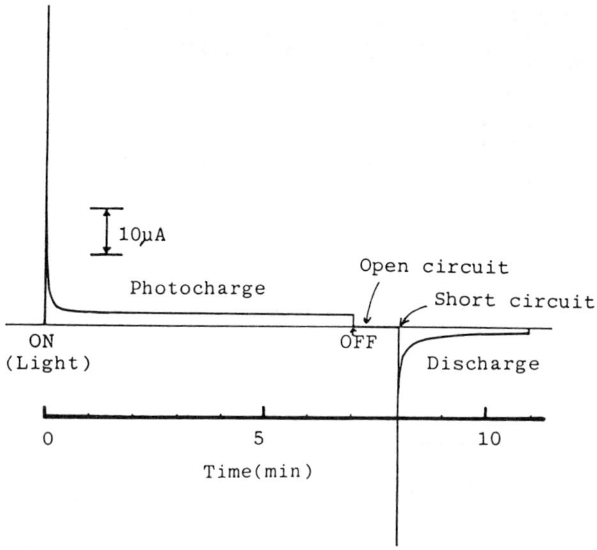

FIGURE 18. Photocharge and dark discharge of the photochargeable device shown in Figure 17.

junction PT membrane coated on an ITO was demonstrated first.[94,95] For a photochargeable device, an ITO electrode was coated with poly(3-methylthiophene) (P3MT) by electropolymerization of 3-methylthiophene present in nitrobenzene (0.1 *M* solution) containing tetrabutylammonium perchlorate (0.05 *M*). Then PB was coated on top of the P3MT membrane by electrodeposition. A photochargeable device was fabricated by sandwiching an aqueous electrolyte solution with a P3MT/PB bilayer-coated ITO and a PB-coated ITO to give the cell composition:[96]

ITO/P3MT/PB/aqueous electrolyte/PB/ITO

This photochargeable device generated photocharging and dark discharging currents induced by switching the irradiation on and off as shown in Figure 19. The *p*-type semiconducting P3MT induced cathodic photocurrent, and resulted in anodic dark discharge current. Since these cycles in this figure were operated under short-circuit conditions, switching the irradiation off immediately induced discharge current. Because the P3MT membrane possesses both semiconducting and ohmic natures, it works also as a discharge electrode in the present device. It should be possible to fabricate various types of photochargeable devices by utilizing semiconductor and charge-storing materials.

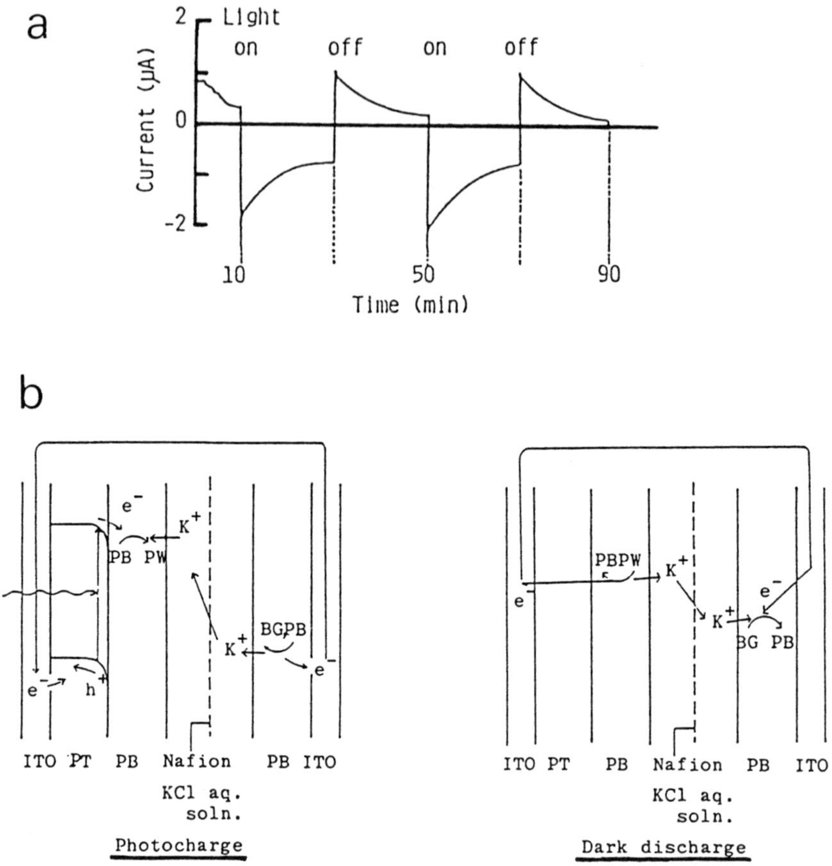

FIGURE 19. (a) Photocharge and dark discharge of the device, ITO/P3MT/PB/aqueous electrolyte/PB/ITO. Electrolyte, 0.2 *M* KCl + 0.01 *M* HCl; electrode area, 1.2 cm^2. (b) Mechanism of photocharge and dark discharge.

B. PHOTOENERGY CONVERSION DEVICES

Photosynthesis in green plants is an important model for constructing artificial photoenergy conversion systems toward future energy resources. This photosynthesis is based on photochemical electron transfer reactions between molecules. In an artificial photosynthetic model, it would also be important to design molecule-based photoprocesses to achieve light-to-chemical energy conversion.[1-8] Utilization of visible light instead of UV light is also a crucial point for the design. As a preliminary model, the present authors have combined a liquid-junction narrow band gap semiconductors with functional molecules to achieve artificial photosynthesis operated in water.

In order to use an n-type narrow band gap semiconductor as a liquid junction device in water, it has to be stabilized against corrosion under irradiation.[97] Because holes formed under irradiation are corrosive in n-type narrow band gap semiconductors, it is important to transport the holes to the water phase before oxidative corrosion of the semiconductor takes place. Inhibition of the semiconductor against water contact is also an important factor because water is a corrosive reactant. The present authors have demonstrated inhibition of corrosion of an n-CdS photoanode (band edge 530 nm) dipped in water by coating it with a hydrophobic polymer membrane incorporating a hole-transporting metal complex as well as a catalyst for oxidation of halide ions or water.[98,99] As for the coating material, polymer-pendant $Ru(bpy)_3^{2+}$ was used in which RuO_2 powders were dispersed as an oxidation catalyst.[99] Stabilization of the CdS photoanode by the coating was shown by disappearance of a couple of redox peaks due to the corrosion product (Cd^{2+}) in the cyclic voltammogram under irradiation. Such a stabilized n-CdS photoanode could oxidize halide ions under visible light irradiation to give halogens, and at the same time hydrogen (H_2) evolved at the Pt counter electrode. The overall reaction is therefore photolysis of hydrogen halide (HX) to produce H_2 and X_2:

$$2HX \xrightarrow{h\upsilon} H_2 + X_2 \tag{6}$$

In the absence of halide ions, water was oxidized on the modified CdS to result in overall water photolysis to produce H_2 and O_2:

$$2H_2O \xrightarrow{h\upsilon} H_2 + O_2 \tag{7}$$

In this photoenergy conversion system, the Ru complex works as a mediator to transport holes from the CdS surface to RuO_2 where catalytic oxidation of X^- or H_2O takes place.

To apply a more active catalyst for water oxidation, various metal complexes were studied as an O_2-evolving catalyst by water oxidation especially under heterogeneous conditions.[100-105] The trinuclear Ru complex:

$$[(NH_3)_5Ru\text{-}O\text{-}Ru(NH_3)_4\text{-}O\text{-}Ru(NH_3)_5]^{6+} \qquad (\mathbf{10}\text{, called Ru-red})$$

active both in homogeneous and heterogeneous conditions, was adsorbed into a Nafion membrane coated on an n-CdS photoanode. This complex worked as both hole mediator and catalyst, and the modified CdS in combination with a Pt counter electrode photolyzed water to give H_2 and O_2 (Figure 20). The H_2 and O_2 yields based on total incident 496 nm monochromatic light were 16%.[105]

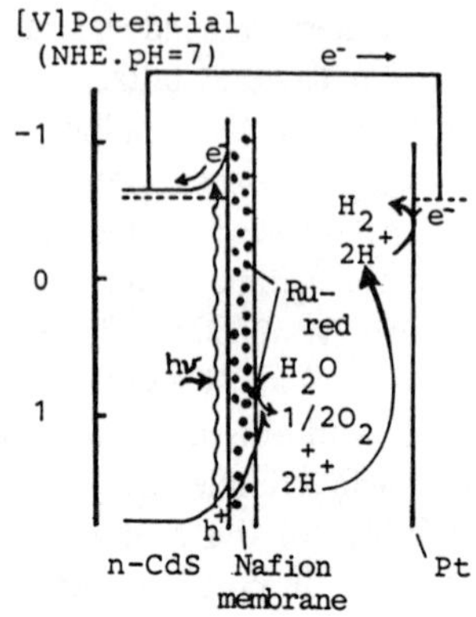

FIGURE 20. Schematic diagram of water photolysis at polymer-modified *n*-CdS (at pH 7).

These photoenergy conversion devices are designed based on molecular catalysis. In the future, semiconductors should be replaced by a photoexcitation center based on a light-absorbing compound to achieve artificial photosynthesis. In order to use a molecule-based photochemical charge separation system instead of using a semiconductor, designs based on kinetics of the electron transfer are of importance in overcoming an energy-consuming back electron transfer.

C. PHOTORESPONSIVE ELECTRODE POSSESSING SIMULTANEOUS OHMIC CHARACTERISTICS

When a semiconductor possesses simultaneous ohmic characteristics, its liquid junction photoelectrode shows an interesting behavior under irradiation. When RuO_2 powders were first put on the surface of a CdS photoanode and the polymer-pendant $Ru(bpy)_3^{2+}$ was coated over them, the surface of CdS/RuO_2 showed a partially ohmic character.[106] The coated cationic Ru complex membrane adsorbed $Fe(CN)_6^{3-/4-}$ from its aqueous solution by electrostatic binding, and the adsorbed $Fe(CN)_6^{3-/4-}$ gave an ohmic response at the CdS electrode superimposed on anodic photocurrent under cyclic potential sweep conditions (Figure 21).

ZnPc dispersed in a polymer membrane also possesses both semiconducting and ohmic characteristics. Figure 22 illustrates typical cyclic voltammograms in the dark and under irradiation of a modified ITO/PVCz membrane containing dispersed ZnPc in contact with a redox electrolyte aqueous solution comprising 10 m*M* MV^{2+} in 0.5 *M* KNO_3. Reversible redox waves due to MV^{2+} were observed superimposed on the cathodic photocurrent induced by the liquid junction *p*-type semiconducting ZnPc.

Although application of this kind of system remains a problem, the interesting new phenomenon of molecular reaction superimposed on photocurrent might find application in some new devices.

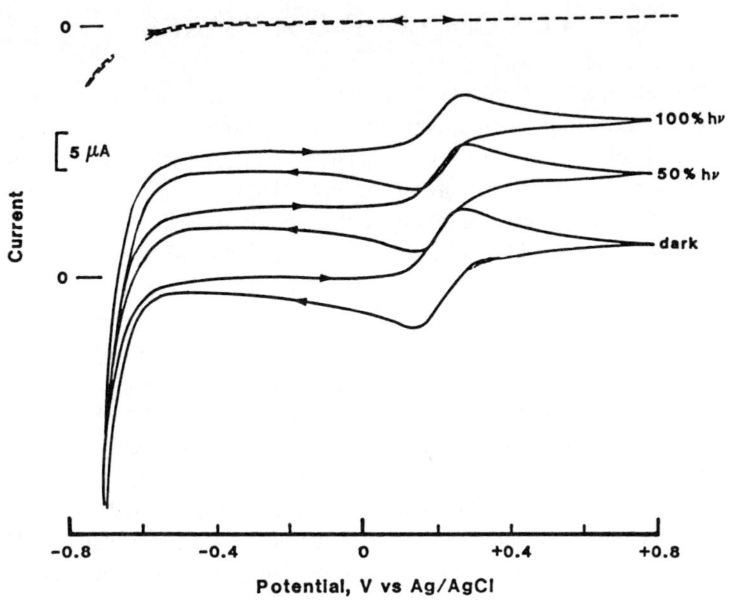

FIGURE 21. Cyclic voltammograms (scan rate: 0.10 V/s) for the **5**-coated *n*-CdS in the dark under illumination dipped in 50 m*M* $Fe(CN)_6^{3-/4-}$ and 1 *M* KNO_3 of solution. The voltammogram for a bare *n*-CdS electrode is shown for comparison (dashed curve, in the dark).

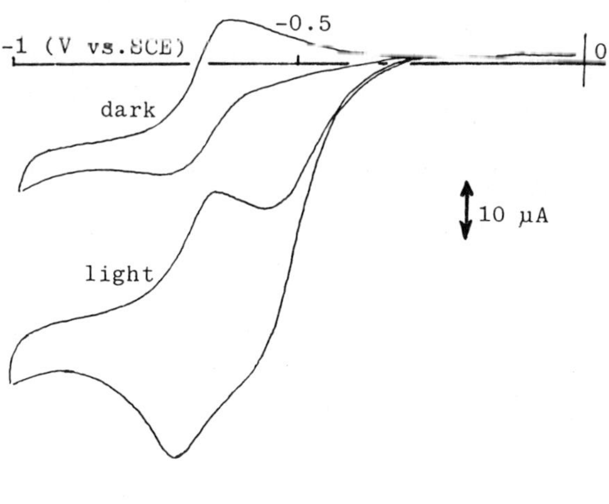

FIGURE 22. Cyclic voltammogram of ITO/PVCz[ZnPc] dipped in 10 m*M* MV^{2+}/0.5 *M* KNO_3 aqueous solution.

VI. CONCLUDING REMARKS

Photoelectrochemical devices utilizing molecule-based excited state reactions are attracting much attention for future photonics as well as for photochemical solar energy conversion systems. Polymer-coated electrodes are at present one of the most promising candidates because of their ease of fabrication and integration of photoreaction components. In order to construct molecule-based photoelectrochemical devices comprising polymer membranes, molecular design and arrangement of polymers incorporating photoreaction components are the key point (as seen in this chapter).

The methodology described here could offer a promising prospect of building photoelectrochemical devices based on molecular excited state reactions at polymer-coated electrodes. A considerable number of applications such as devices of photoresponse, photocharging, photoenergy conversion, and photoelectrochemical sensors are attracting a great deal of attention. Considering the great advantages of molecule-modified electrodes by utilizing polymeric compounds as a first choice, the authors believe that new fields of science and industry would be possible. This hope could be realized by organized collaboration between research institutes, universities, and industries.

REFERENCES

1. **Grätzel, M.,** *Heterogeneous Photochemical Electron Transfer,* CRC Press, Boca Raton, FL, 1989.
2. **Grätzel, M., Ed.,** *Energy Resources through Photochemistry and Catalysis,* Academic Press, New York, 1983.
3. **Jame, R. W., Jr. and Dan, M., Eds.,** *Photochemical Energy Conversion,* Elsevier, New York, 1989.
4. **Pelizzetti, E. and Schiavello, M., Eds.,** *Photochemical Conversion and Storage of Solar Energy,* Kluwer Academic Pub., Dordrecht, The Netherlands, 1991.
5. **Kaneko, M. and Yamada, A.,** Solar energy conversion by functional polymers, *Adv. Polym. Sci.,* 55, 1, 1984.
6. **Kaneko, M. and Wöhrle, D.,** Polymer-coated electrodes: new materials for science and industry, *Adv. Polym. Sci.,* 84, 141, 1988.
7. **Grätzel, M.,** Photo-induced water splitting in heterogeneous solution, in *Photochemical Conversion and Storage of Solar Energy,* Connolly, J. S., Ed., Academic Press, New York, 1981, 131.
8. **Lehn, J. M.,** Photoinduced generation of hydrogen and oxygen from water, in *Photochemical Conversion and Storage of Solar Energy,* Connolly, J. S., Ed., Academic Press, New York, 1981, 161.
9. **Möbius, D., Ed.,** *Langmuir-Blodgett Films 3,* Vols. 1 and 2, Elsevier, London, 1988.
10. **Yamazaki, I., Tamai, N., Yamazaki, T., Murakami, A., Mimuro, M., and Fujita, Y.,** Sequential excitation energy transport in stacking multilayers: comparative study between photosynthetic antenna and Langmuir-Blodgett multilayers, *J. Phys. Chem.,* 92, 5035, 1988.

11. **Yamazaki, I., Tamai, N., and Yamazaki, T.,** Electronic excitation transfer in organized molecular assemblies, *J. Phys. Chem.*, 94, 516, 1990.
12. **Fujihira, M. and Yamada, H.,** Molecular photodiodes consisting of unidirectionally oriented amphipathic acceptor-sensitizer-donor triads, *Thin Solid Films,* 160, 125, 1988.
13. **Kondo, T., Yamada, H., Nishiyama, K., Suga, K., and Fujihira, M.,** Electron transfer quenching of an excited amphiphilic $Ru(bpy)_3^{2+}$ derivative in Langmuir-Blodgett films containing various ferrocene derivatives as electron donors, *Thin Solid Films,* 179, 463, 1989.
14. **Fujihira, M., Sakomura, M., and Kamei, T.,** Simulation of the primary process in photosynthetic reaction centre by monolayer with charge separation triad and antenna molecules, *Thin Solid Films,* 180, 43, 1989.
15. **Kaneko, M.,** Photochemistry of polymer-metal-complexes, *Kobunshi,* 40, 648, 1991.
16. **Turro, N. J.,** *Modern Molecular Photochemistry,* Benjamin/Cummings, Menlo Park, CA, 1978.
17. **Nocera, D. G., Winkler, J. R., Yocom, K. M., Bordignon, E., and Gray, H. B.,** Kinetics of intermolecular electron transfer from Ru^{II} to Fe^{III} in Ruthenium-modified cytochrome c, *J. Am. Chem. Soc.*, 106, 5145, 1984.
18. **Li, T. T.-T. and Weaver, M. J.,** Intramolecular electron transfer at metal surfaces. IV. Dependence of tunneling probability upon donor-acceptor separation distance, *J. Am. Chem. Soc.*, 106, 6107, 1984.
19. **Isied, S. S., Kuehn, C., and Worosila, G.,** Ruthenium-modified cytochrome c: temperature dependence of the rate of intramolecular electron transfer, *J. Am. Chem. Soc.*, 106, 1722, 1984.
20. **Miller, J. R., Calcaterra, L. T., and Closs, G. L.,** Intramolecular long-distance electron transfer in radical anions: the effects of free energy and solvent on the reaction rates, *J. Am. Chem. Soc.*, 106, 3047, 1984.
21. **Whitten, D. G., Russell, J. C., and Schmehl, R. H.,** Photochemical reactions in organized assemblies: environmental effects on reaction occurring in micelles, vesicles, films and multilayer assemblies and at interfaces, *Tetrahedron,* 38, 2455, 1982.
22. **Fendler, J. H.,** Microemulsions, micelles and vesicles as media for membrane mimetic photochemistry, *J. Phys. Chem.*, 84, 1485, 1980.
23. **Ford, W. E. and Tollin, G.,** Chlorophyll photosensitized electron transfer in phospholipid vesicle bilayers: inside vs. outside asymmetry, *Photochem. Photobiol.*, 36, 647, 1982.
24. **Woodle, M. C. and Mauzerall, D.,** Photoinitiated movements in bilayer membranes containing magnesium octaethylporphyrin, *Biophys. J.*, 50, 431, 1986.
25. **Gurp, M., Ginkel, G., and Levien, Y. K.,** On the influence of the lipid matrix on energy transfer processes in self-assembling chlorophyll-lipid systems, *Biochim. Biophys. Acta,* 938, 71, 1988.
26. **Grossweiner, L. I. and Goyal, G. C.,** Photosensitization of liposomes by porphyrins, *J. Photochem.*, 25, 253, 1984.
27. **Calvin, M.,** Simulating photosynthetic quantum conversion, *Acc. Chem. Res.*, 11, 369, 1978.
28. **Tsuchida, E., Kaneko, M., Nishide, H., and Hoshino, M.,** Distance dependence of electron transfer from liposome-embedded (alkanephosphocholine-porphinato)zinc, *J. Phys. Chem.*, 90, 2283, 1986.
29. **Kaneko, M., Tsuchida, E., and Imai, Y.,** Incorporation and distribution of benzoquinone into liposomes containing amphilic zinc(II) porphyrin as studied by the quenching of the excited state, *J. Chem. Soc. Faraday Trans.*, 87, 83, 1991.
30. **Luque, A., Sala, G., Palz, W., Santos, G. D., and Helm, P., Eds.,** *Proc. 10th Eur. Photovoltaic Solar Energy Conf.*, Kluwer Academic Publ., Dordrecht, The Netherlands, 1991.
31. **Rabinowitch, E.,** The photogalvanic effect. I. The photochemical properties of the thionine-iron system, *J. Chem. Phys.*, 8, 551, 1940.

32. **Rabinowitch, E.**, The photogalvanic effect. II. The photogalvanic properties of the thionine-iron system, *J. Chem. Phys.*, 8, 560, 1940.
33. **Eisenberg, M. and Silverman, H. P.**, Photoelectrochemical cells, *Electrochim. Acta*, 5, 1, 1961.
34. **Yoshida, M. and Oshida, I.**, Photogalvanic cells by the use of ion-exchange membranes, *Jpn. Appl. Phys.*, 33, 34, 1964.
35. **Kaneko, M. and Yamada, A.**, Photoinduced potentials by photoredox reaction systems and the addition effects of EDTA, *Rep. Inst. Phys. Chem. Res.*, 52, 210, 1976.
36. **Kaneko, M. and Yamada, A.**, Photopotential and photocurrent induced by tolusafranine-ethylenediaminetetraacetic acid system, *J. Phys. Chem.*, 81, 1213, 1977.
37. **Shigehara, K., Nishimura, M., and Tsuchida, E.**, Photo-induced electricity generated by thin-layer photogalvanic cells containing thionine and iron(II) salt, *Bull. Chem. Soc. Jpn.*, 50, 3397, 1977.
38. **Albery, W. J., Foulds, A. W., Hall, K. J., Hillman, A. R., Egdell, R. G., and Orchard, A. F.**, Thionine coated-electrode for photogalvanic cells, *Nature (London)*, 282, 793, 1979.
39. **Albery, W. J., Foulds, A. W., Hall, K. J., and Hillman, A. R.**, The thionine-coated electrode for photogalvanic cells, *J. Electrochem. Soc.*, 127, 654, 1980.
40. **Tamilarasan, R. and Natarajan, P.**, Photovoltaic conversion by macromolecular thionine films, *Nature (London)*, 292, 224, 1981.
41. **Kamat, P. V.**, Photoelectrochemistry in colloidal systems: interfacial electron transfer between colloidal titanium dioxide and thionine in acetontrile, *J. Photochem.*, 28, 513, 1985.
42. **Kamat, P. V.**, Photoelectrochemistry in colloidal systems. II. A photogalvanic cell based on titanium oxide semiconductor colloid, *J. Chem. Soc. Faraday Trans. 1*, 81, 509, 1985.
43. **Quickenden, T. I. and Harrison, I. R.**, The power conversion efficiency of the thionine photoelectrochemical cell containing uncoated and thionine coated photoelectrodes, *J. Electrochem. Soc.*, 132, 81, 1985.
44. **Tamilarasan, R., Ramaraj, R., Subramanian, R., and Natarajan, P.**, Light-induced electron-transfer reactions at electrodes coated with macromolecular thionine and ruthenium systems, *J. Chem. Soc. Faraday Trans. 1*, 80, 2405, 1984.
45. **Ramaraj, R., Tamilarasan, R., and Natarajan, P.**, Luminescence quenching and flash photolysis studies of acrylamidomethylthionine copolymers, *J. Chem. Soc. Faraday Trans. 1*, 81, 2763, 1985.
46. **Kaneko, M., Ochiai, M., and Yamada, A.**, Photoresponsive carbon electrode coated with polyanion layer containing the tris(2,2′-bipyridyl)ruthenium(II) complex, *Makromol. Chem. Rapid Commun.*, 3, 299, 1982.
47. **Oyama, N., Yamaguchi, S., Kaneko, M., and Yamada, A.**, Photocurrent generation of graphite electrodes coated with polymer films containing *tris*(2,2′-bipyridyl)-ruthenium(II) and viologen, *J. Electroanal. Chem.*, 139, 215, 1982.
48. **Kaneko, M., Yamada, A., Oyama, N., and Yamaguchi, S.**, Photoresponsive graphite electrode coated with a film of polymer with pendant *tris*(2,2′-bipyridyl)ruthenium(II) complex, *Makromol. Chem. Rapid Commun.*, 3, 769, 1982.
49. **Kaneko, M., Yamada, A., and Oyama, N.**, Electron-transfer reaction and formation of active species induced by both applied potential and photoexcitation, *Nippon Kagaku Kaishi*, 1984, 1810, 1984.
50. **Kaneko, M., Moriya, S., Yamada, A., Yamamoto, H., and Oyama, N.**, Photoresponsive electrode coated with bilayer membranes of polymer pendant $Ru(bpy)_3^{2+}$ complex and viologen, *Electrochim. Acta*, 29, 115, 1984.
51. **Ohsaka, T., Yamamoto, H., Kaneko, M., Yamada, A., Nakamura, M., Nakamura, S., and Oyama, N.**, Electrode kinetics of viologen polymers attached on graphite surfaces, *Bull. Chem. Soc. Jpn.*, 57, 1844, 1984.

52. **Ohsaka, T., Oyama, N., Sato, K., and Matsuda, H.,** The influence of supporting electrolyte on kinetics of electroreduction process of poly(methylviologen), poly(methylviologen)-Nafion complex, and poly(methylviologen)-poly(*p*-styrenesulfonate)complex coated on electrodes, *J. Electrochem. Soc.*, 132, 1871, 1985.
53. **Oyama, N., Ohsaka, T., Yamamoto, H., and Kaneko, M.,** Charge-transfer reactions in pendant viologen polymers coated on graphite electrodes and at electrode/pendant viologen polymer film interfaces, *J. Phys. Chem.*, 90, 3850, 1986.
54. **Ohsaka, T., Yamamoto, H., and Oyama, N.,** Thermodynamic parameters for charge-transfer reactions in pendant viologen polymers coated on graphite electrodes and at electrode/pendant viologen polymer film interfaces, *J. Phys. Chem.*, 91, 3775, 1987.
55. **Raman, C. V.,** On the mean distance between neighbouring molecules in a fluid, *Philos. Mag.*, 47, 671, 1924.
56. **Gans, R.,** Über den Abstand der molekeln, *Phys. Z.*, 23, 108, 1922.
57. **Kaneko, M., Hou, X.-H., and Yamada, A.,** Specific quenching of the photoexcited state of *tris*(2,2′-bipyridine)ruthenium(II) by colloidal Prussian blue, *J. Chem. Soc. Faraday Trans. 1*, 82, 1637, 1986.
58. **Nagai, K., Takamiya, N., and Kaneko, M.,** Photoluminescence as a probe for molecular dispersion in solid matrix, unpublished data, 1992.
59. **Yilgor, I. and McGrath, D. E.,** Polysiloxane-containing copolymers: a survey of recent developments, *Adv. Polym. Sci.*, 86, 1, 1988.
60. **Yamada, K., Ueno, Y., Ikeda, K., Takamiya, N., and Kaneko, M.,** Synthesis of polysiloxanes with pendant viologen groups and their characteristic behavior of aggregation and deposition observed on electron transfer, *Makromol. Chem.*, 191, 2871, 1990.
61. **Ueno, Y., Yamada, K., Yokota, T., Ikeda, K., Takamiya, N., and Kaneko, M.,** Photoresponsive electrode based on the reaction between oxygen and excited $Ru(bpy)_3^{2+}$ complex incorporated into a coated Nafion film, *Inorg. Chim. Acta*, in press.
62. **Kaneko, M., Takabayashi, N., and Yamada, A.,** Simultaneous evolution of hydrogen and oxygen by water photolysis with Prussian blue and *tris*(2,2′-bipyridyl)ruthenium(II) complex, *Chem. Lett.*, 1647, 1982.
63. **Kaneko, M., Takabayashi, N., Yamauchi, Y., and Yamada, A.,** Water photolysis by means of visible light with a system composed of Prussian blue and the *tris*(2,2′-bipyridine)ruthenium(II) complex, *Bull. Chem. Soc. Jpn.*, 57, 156, 1984.
64. **Toshima, N., Lin, R.-J., and Kaneko, M.,** A new preparation method of electrochromic Prussian blue films by casting a solution and their electrochemical properties, *Chem. Lett.*, 485, 190, 1982.
65. **Lin, R.-J. and Toshima, N.,** Spectroelectrochemistry of new Prussian blue films prepared by a cast method, *Bull. Chem. Soc. Jpn.*, 64, 136, 1991.
66. **Itaya, K., Ataka, T., and Toshima, S.,** Spectroelectrochemistry and electrochemical preparation method of Prussian blue modified electrodes, *J. Am. Chem. Soc.*, 104, 4767, 1982.
67. **Itaya, K., Uchida, I., and Neff, V. D.,** Electrochemistry of polynuclear transition metal cyanides: Prussian blue and its analogues, *Acc. Chem. Res.*, 19, 162, 1986.
68. **Busser, H. J., Schwarzenbach, D., Petter, W., and Ludi, A.,** The crystal structure of Prussian blue: $Fe_4[Fe(CN)_6]^{3-} \cdot xH_2O$, *Inorg. Chem.*, 16, 2704, 1977.
69. **Kaneko, M.,** Photosensing device composed of bilayer membranes of polymer-pendant $Ru(bpy)_3^{2+}$ and Prussian blue, *J. Macromol. Sci. Chem.*, A24, 357, 1987.
70. **Kaneko, M., Hara, S., and Yamada, A.,** A photoresponsive graphite electrode coated with Prussian blue, *J. Electroanal. Chem.*, 194, 165, 1985.
71. **Kaneko, M. and Yamada, A.,** Photoresponsive graphite electrode coated with bilayer membranes of Prussian blue and polymer-pendant $Ru(bpy)_3^{2+}$, *Electrochim. Acta*, 31, 273, 1986.
72. **Leempoel, P., Fan, F.-R. F., and Bard, A. J.,** Semiconductor electrodes. 50: effect of mode of illumination and doping on photochemical behavior of phthalocyanine films, *J. Phys. Chem.*, 87, 2948, 1983.

73. **Giraudeau, A., Fan, F.-R. F., and Bard, A. J.,** Semiconductor electrode. 30: spectral sensitization of the semiconductors in *n*-TiO_2 and *n*-WO_3 with metal phthalocyanines, *J. Am. Chem. Soc.*, 102, 5137, 1980.
74. **Rieke, P. C. and Armstrong, N. R.,** Light-assisted aqueous redox reactions at chlorogallium phthalocyanine thin-film photoconductors: dependence of the photopotential on the formal potential of the redox couple and evidence for photoassisted hydrogen evolution, *J. Am. Chem. Soc.*, 106, 47, 1984.
75. **Klofta, T. J., Rieke, P. C., Linlous, C. A., Buttner, W. J., Nanthakumar, A., Newborn, T. D., and Armstrong, N. R.,** Tri- and tetravalent phthalocyanine thin film photoelectrodes: comparison with other metal and demetallated phthalocyanine systems, *J. Electrochem. Soc.*, 132, 2134, 1985.
76. **Loutfy, R. O. and McIntyre, L. F.,** Phthalocyanines: electrolyte Schottky-junction devices, *Can. J. Chem.*, 61, 72, 1983.
77. **Minami, N., Sasaki, K., and Tsuda, K.,** Improvement of the performance of particulate phthalocyanine photovoltaic cells by the uses of polar polymer binders, *J. Appl. Phys.*, 54, 6764, 1983.
78. **Yamada, K., Ueno, Y., Ikeda, K., Takamiya, N., and Kaneko, M.,** Photoresponsive electrode coated with bilayer membranes of polysiloxane-viologen and phthalocyanine, *Denki Kagaku*, 57, 1129, 1989.
79. **Ishiji, T. and Kaneko, M.,** Luminescence characteristics of *tris*(2,2′-bipyridine)ruthenium(II) complex adsorbed in a Nafion film, *Kobunshi Ronbunshu*, 47, 869, 1990.
80. **Kaneko, M., Wöhrle, D., Schlettwein, D., and Schmidt, V.,** Dioxygen sensitivity of a photoexcited thin film of phthalocyanine dispersed in poly(*N*-vinylcarbazole), *Makromol. Chem.*, 189, 2419, 1988.
81. **Wöhrle, D., Schlettwein, D., Kirschenmann, M., Kaneko, M., and Yamada, A.,** The combination of phthalocyanines and polymers for electrochemically or photoelectrochemically induced processes, *J. Macromol. Sci. Chem.*, A27, 1239, 1990.
82. **Schlettwein, D., Kaneko, M., Yamada, A., Wöhrle, D., and Jaeger, N. I.,** Light-induced dioxygen reduction at thin film electrodes of various porphyrins, *J. Phys. Chem.*, 95, 1748, 1991.
83. **Kaneko, M. and Wöhrle, D.,** Novel photogalvanic effect of the viologen cation radical formed electrochemically in the presence of oxygen, *J. Electroanal. Chem.*, 307, 209, 1991.
84. **Manassen, J., Hodes, G., and Cahen, D.,** Photoelectrochemical energy conversion and storage: the polycrystalline CdSe cell with different storage modes, *J. Electrochem. Soc.*, 124, 532, 1977.
85. **Yonezawa, Y., Okai, M., Ishino, M., and Hada, H.,** A photochemical storage battery with an *n*-GaP photoelectrode, *Bull. Chem. Soc. Jpn.*, 56, 2873, 1983.
86. **Kanda, S. and Murray, R. W.,** Optical energy storage by bilayer electrodes: polymerized coordination compounds of ruthenium or iron, *Bull. Chem. Soc. Jpn.*, 58, 3010, 1985.
87. **Betz, G., Tributsch, H., and Fiechter, S.,** Photointercalation and optical information storage using $Cu_{6-x}PS_5I$, *J. Electrochem. Soc.*, 131, 640, 1984.
88. **Gerriten, H. J., Ruppel, M., and Wurfel, P.,** A photoelectrochemical storage cell with *n*-CdSe and *p*-CdTe electrodes, *J. Electrochem. Soc.*, 131, 2037, 1984.
89. **Neff, V. D.,** Some performance characteristics of a Prussian blue battery, *J. Electrochem. Soc.*, 132, 1382, 1985.
90. **Kaneko, M.,** Polynuclear-metal-complex battery, *J. Polym. Sci. Polym. Lett. Ed.*, 24, 435, 1986.
91. **Konda, K., Ochiai, J., and Hayashi, H.,** Polymerization of transition metal complexes in solid polymer electrolytes, *J. Chem. Soc. Chem. Commun.*, 168, 1986.
92. **Kaneko, M. and Okada, T.,** A secondary battery composed of multilayer Prussian blue and its reaction characteristics, *J. Electroanal. Chem.*, 255, 45, 1988.

93. **Kaneko, M., Okada, T., Minoura, H., Sugiura, T., and Ueno, Y.,** Photochargeable multilayer membrane device composed of CdS film and Prussian blue battery, *Electrochim. Acta,* 35, 291, 1990.
94. **Kenmochi, T., Tsuchida, E., Kaneko, M., and Yamada, A.,** Photoelectrochemical response of liquid junction poly(thienylene), *Electrochim. Acta,* 30, 1405, 1985.
95. **Kaneko, M., Yamada, A., Kenmochi, T., and Tsuchida, E.,** Photoresponse of a Schottky junction polythienylene film, *J. Polym. Sci. Polym. Lett. Ed.,* 23, 629, 1985.
96. **Kaneko, M., Takahashi, K., and Tsuchida, E.,** A photochargeable device composed of bilayer membranes of poly(3-methylthiophene) and Prussian blue, *J. Electroanal. Chem.,* 227, 255, 1987.
97. **Kaneko, M., Okada, K., Teratani, S., and Taya, K.,** Visible light cleavage of water by CdS photoanode coated with polymer-pendant $Ru(bpy)_3^{2+}$ film containing RuO_2 dispersions, *Electrochim. Acta,* 32, 1405, 1987.
98. **Rajeshwar, K., Kaneko, M., and Yamada, A.,** Regenerative photoelectrochemical cells using polymer-coated *n*-GaAs photoanodes in contact with aqueous electrolytes, *J. Electrochem. Soc.,* 130, 38, 1983.
99. **Rajeshwar, K., Kaneko, M., Yamada, A., and Noufi, R. N.,** Photoelectrochemical oxidation of halide ions at naked, catalytically modified, and polymer-coated *n*-CdS electrodes in aqueous media, *J. Phys. Chem.,* 89, 806, 1985.
100. **Ramaraj, R., Kira, A., and Kaneko, M.,** Oxygen evolution by water oxidation mediated by heterogeneous manganese complexes, *Angew. Chem. Int. Ed. Engl.,* 25, 825, 1986.
101. **Ramaraj, R., Kira, A., and Kaneko, M.,** O_2-generation by oxidation of water with di- and trinuclear ruthenium complexes as homogeneous and heterogeneous catalyst, *Angew. Chem. Int. Ed. Engl.,* 25, 1009, 1986.
102. **Ramaraj, R., Kira, A., and Kaneko, M.,** Electrochemistry and stability studies of oxo-bridged dinuclear ruthenium(III) complexes for water oxidation, *J. Chem. Soc. Faraday Trans. 1,* 82, 3515, 1986.
103. **Ramaraj, R., Kira, A., and Kaneko, M.,** Oxygen evolution by water oxidation with polynuclear ruthenium complexes, *J. Chem. Soc. Faraday Trans. 1,* 83, 1539, 1987.
104. **Yao, G.-J., Kira, A., and Kaneko, M.,** Enhanced oxygen evolution through electrochemical water oxidation mediated by polynuclear complexes embedded in a polymer film, *J. Chem. Soc. Faraday Trans. 1,* 84, 4451, 1988.
105. **Kaneko, M., Yao, G.-J., and Kira, A.,** Efficient water cleavage with visible light by a system mimicking photosystem II, *J. Chem. Soc. Chem. Commun.,* 1338, 1989.
106. **Rajeshwar, K. and Kaneko, M.,** Is the *n*-CdS/RuO_2 interface ohmic or is it a Schottky barrier, *J. Phys. Chem.,* 89, 3587, 1985.

INDEX

D

E

F

G

H

I

K

L

M

N

Q

R

S